HENNING ANDERSEN

Active Arithmetic!

Movement and Mathematics Teaching in the Lower Grades of a Waldorf School

Published by:

Waldorf Publications
Research Institute for Waldorf Education
38 Main Street
Chatham, New York 12037

Active Arithmetic! (Regn og Hop!) first published by
Brage Forlagstrykkeri, Denmark

Second edition © 2014
ISBN 978-1-936367-50-4
First edition 1984, 1995, 2001, 2013
ISBN 978-87-88258-74-5

Author and illustrator: Henning Andersen
English translation: Archie Duncanson and Verner Pedersen
Editor and designer, first edition: David Mitchell
Layout, second edition: Ann Erwin
Proofreader, copy editor: Ann Erwin

Illustrations: pages 43–44: Albrecht Dürer; page 38: "The Lion," from *Stray Leaves from Strange Literature*, a book of stories reconstructed by Lafcadio Hearn

All rights reserved. No part of this publication may be reproduced or translated, in any form or by any means, without permission from the publisher, except for short quotations for reviews.

Curriculum Series

The Publications Committee of the Research Institute is pleased to bring forward this publication as part of its Curriculum Series. The thoughts and ideas represented herein are solely those of the author and do not necessarily represent any implied criteria set by Waldorf Publications. It is our intention to stimulate as much writing and thinking as possible about our curriculum, including diverse views. Please contact us at patrice@waldorf-research.org with feedback on this publication as well as requests for future work.

CONTENTS

	Translators' Foreword	4
	Foreword	5
1	Introduction: On Teaching Arithmetic to Children	7
2	The Threefold Nature of Man	13
3	Number Qualities and Hidden Secrets	19
4	Rhythm and Numbers	46
5	Multiplication Tables	64
6	More Tables: The Relations between Numbers	76
7	Drawing and Arithmetic	95
8	Numbers as Quantities	116
9	Concluding Remarks on Mathematics, Memory and Games	165

TRANSLATORS' FOREWORD

At the beginning of this book, there is a discussion of the educational system in Denmark. It may be argued that some of this description could have been omitted from an English edition. The situation in Denmark, however, is perhaps a reflection of what is happening in the rest of the world, so we have included the Danish experience.

It was our privilege to be colleagues of the author for several years. During work on the translation, many of our earlier conversations on the teaching of arithmetic came back to mind. Henning was always so very enthusiastic and provided us understanding of both the practical exercises and their philosophical background.

After such conversations, arithmetic lessons became a real joy for both teacher and children, and we have no doubt that any teacher studying this book will have the same experience.

Verner Pedersen
Archie Duncanson

FOREWORD

The aim of this book is not to produce a complete account of the Steiner/Waldorf method of teaching arithmetic in the younger classes. This may be evident by the fact that the important areas of the four arithmetic operations are mentioned but not treated in depth, that the introduction of numbers is not touched upon, and that written arithmetic is mentioned only superficially. In addition, other important subjects have been left out altogether.

The exclusive purpose is to examine those parts of the teaching of arithmetic that can be developed through movement and which can satisfy the child's need to experience soul qualities in everything he or she does. Many may feel that the latter sounds over ambitious, especially in relation to such a "dry" subject as arithmetic.

The subject certainly has many "dry" aspects which must be considered, but only at the age when the child has developed sufficiently. It is therefore not simply a question of skills and know-how, but above all of judging the need of a particular age.

The fundamental views expressed in this book are (1) that during the first years of school, the will and the emotional life of the child play bigger roles than the intellect, and (2) that this relationship does not exclude the child from appreciating a subject such as arithmetic. On the contrary, these aspects of the human soul-life can find their very best ally in arithmetic.

The book is addressed not only to teachers of younger classes, but also to parents and other adults who work daily with children. They might be trying to satisfy the children's need for movement with a game, or with rhythmical activity in bigger or smaller groups, or perhaps they might want to provide a quieter occupation around the family dining table. In all of these situations, a child's need can be satisfied by the exercises which follow, and at the same time, they will provide important preparation for more intellectual activity later on.

The author has taught in a Rudolf Steiner school for many years and while working on the text has held certain children and certain classes in mind. He invites readers to do the same. In other words, one should never apply recipes directly—neither in the art of cooking, nor in the art of teaching—but modify them so that they are in tune with the individuals you wish to nourish. Only then will they have value for both body and soul.

Henning Andersen

Chapter 1
INTRODUCTION
ON TEACHING ARITHMETIC TO CHILDREN

Those readers who are getting on in life will remember from their schooldays at least one horribly boring textbook which they had to take out of their bag each day for arduous work. From the first page to the last page, it was filled with numbers and divided into lessons for the whole year: Lesson 1, Lesson 2, Lesson 3, etc. Without exception each lesson was divided into oral exercises and written exercises, and these were divided in turn into number exercises and written problems. Each lesson contained a multiplication table, exercises in addition, then perhaps a long section on subtraction, and later, exercises in multiplication and division.

There was a teacher, too. He or she had the task of reaching the last page of the book by the end of the year, and this did not leave him or her much scope for taking initiative in teaching.

The author remembers his own teacher in this way: He came in with the book in his hand, went at once to the blackboard, found a suitable piece of chalk, and wrote in the middle of the board the number 7. He then looked over the top of his spectacles at his 30 pupils (in order to ensure that they understood that today's lesson was the sevens table), whereupon he wrote, preceding the number 7, a multiplication sign, and in front of this, a pair of brackets, and finally, within the brackets, a series of random numbers. All this he managed to do with his right hand while facing the class.

His head was turned in such a way that he resembled an Egyptian statue, if one ignores the change of fashion and the use of spectacles. About one third of the class was tested on the multiplication table: 4 x 7 = 28, 9 x 7 = 63, 1 x 7 = 7, etc. The results were good. Each pupil was tested several times a week.

We liked our teacher. We also feared him a little and wanted to please him. His teaching was not exciting (there was no thought of that), but we came to like him as we got to know him. Now and then he undoubtedly dreamed of a different arithmetic book, but each day he woke to the realities of what his school had to offer him.

We were tired and stiff when the lesson was over, and we stretched our limbs on the way out. The only exercise we had during the lesson was when we had to stand up to recite the table. The arithmetic book itself did not require anything but mental achievement, as it did not encourage physical activity nor contain any pictures that might perhaps have aroused interest—it was as dead as the telephone directory.

All this has changed since then. The introduction of modern mathematics together with interest in the creative subjects has changed the character of the teaching of arithmetic. The new attitude can be seen the moment we compare our childhood arithmetic book with one from the 1960s. Here there are drawings and colors. In the written problems, there is not only talk about trucks that are to be loaded with sand to be transferred into circular and rectangular shaped boxes, but everything is intensified, and so too is the activity.

In the 1960s, experience and activity were what it all was about. "At each stage we must ensure that we look at the curriculum in the light of the need for activity and experience. It must always be the task of the school—along with imparting basic, elementary knowledge—to develop the children's fundamental human abilities and talents, and to attempt to awaken them to a real understanding of the problems of daily life.

"It is the aim of the school to qualify children to go into the community and employment well prepared to comply with the demands that can be reasonably expected of them. However, it is first and foremost the task of the school to promote opportunities for children to grow up as harmonious, happy and good human beings."

These quotations are from *Educational Guidelines for the Danish School*, in Denmark the so-called "blue paper" from 1960. In the second quotation, the expression "first and foremost" is of great historic interest when one compares it with the wording of previous Danish School Acts and what was said then about the aims of education. Earlier, the aim of education was "first and foremost" to secure elementary knowledge, but now "first and foremost" refers to completely different aspects of human existence.

This is very exciting and interesting!

Likewise, it is interesting to look in the same blue paper under the heading arithmetic and mathematics and read about the aims of the subject: "to impart knowledge and skills to the pupils, to train and exercise the skills in arithmetic that they will need in life outside the school in family, community and industry, and to give the pupils confidence in the basic rules and methods of geometry and arithmetic."

It would be too much to claim that the high ideals set forth with the words "first and foremost" in the introductory chapter have been completely removed by the time we get to the section on arithmetic. It would be true, however, to say that the ideals have diminished somewhat, under the influence of traditional ways of thinking.

What has occurred here is what always happens when something new is forced upon us. We realize with our intellect what we ought to do. We see from experience in our own classroom with our children—and we convince ourselves—that change is necessary, but we do not take in the practical skills which necessarily must accompany our conceptual and emotional conviction. This situation is expressed by Henrik Ibsen in *Per Gynt*, where he lets the main character observe, at a distance, the young woodcutter trying to mutilate his hand with an ax in order to avoid military service:

Yes, think it, wish it, and will it, too.
But do it! No, I do not understand.

It is the execution itself which is difficult. We are clever in developing ideals, and we are good at preaching morals. However, we are often deficient when it comes to exercising our morals, and it is the same when we are to put into practice and actually carry out the teaching of arithmetic in the early years of school.

What can we do in order to create activity and experience in our arithmetic lessons?

The solution is, of course, not to be found in borrowing stimulating elements and motifs from areas where there is no real connection with the subject. We often do this, however. A child naturally finds it amusing to solve a problem of the sort at the top of the next page.

However, the question here is whether the case has been well considered. The children have been given an illustration which belongs

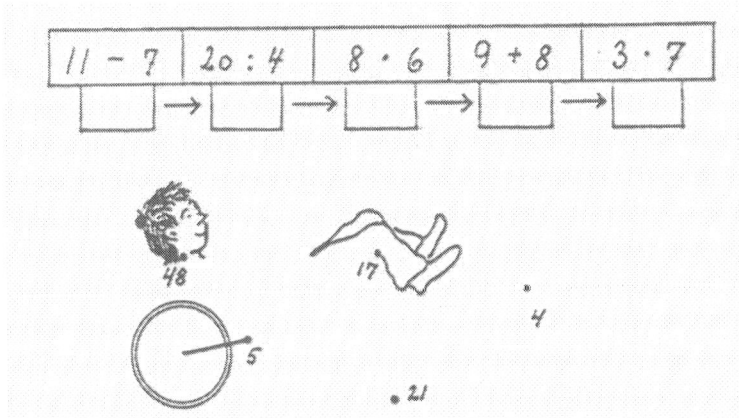

Complete the operation and connect the dots.

somewhere else—often in zoology—and its presence only emphasizes, by the way we have organized things, that arithmetic itself is of no interest. We use horses and elephants to help us because we have not yet found the approach which, being in complete harmony with the nature of the child, contains the motivation within itself.

At the same time something else unfortunate occurs in a different area. The four-footed creatures we meet in our arithmetic books often belong to rather peculiar species, and the gaily-colored butterflies we conjure up for this purpose are all highly unaesthetic. They can never be anything else, and because our children have artistic perception to a degree that we adults often forget, our "carrot-method" can have only a destructive effect. If we study these arithmetic books, we will find that most of them distort the image of people which we want to convey to our children. This happens, for example, when we ask the pupils to pair different-shaped hats (cone, pyramid, sphere, cylinder, etc.) with correspondingly shaped heads, in order to teach the difference between even and odd numbers or other relationships in mathematics. Adults may think that such caricatures are funny, but children may consciously wonder why adults misuse the paper and printer's ink. Besides, the human instrument was not created in this way according to geometrical forms, so geometry has also been grossly misused.

The fundamental question which must be asked in connection with the teaching of arithmetic is the same as for all other subjects: How is it related to the child's nature and the needs of a particular age? From the "blue paper" paragraph on arithmetic, we may not assume that this question has been ignored, but it is clear that the position has been the same as it always has been in relation to arithmetic. The practical side has overwhelmingly dominated—the side which relates to the adult community with its occupations and requisite skills. In one way this is understandable, because arithmetic and mathematics are the two subjects which are most difficult for us to see in relation to the child's world and where it is easiest to make the mistake of regarding children as adults in miniature.

This changes, however, if we begin to look at aspects of human nature other than those which we usually use as adults in doing mathematics—and then realize that these other aspects can also be recognized within arithmetic and mathematics.

Chapter 2
THE THREEFOLD NATURE OF MAN

In the world of education, discussion once centered around the necessity of subjects especially designed for recreation. This has come to an end. The term "recreational subject" implies that children must have lessons which will refresh them from other subjects—they need to rest after the strain of the real work.

However, we have stopped thinking in this way, and instead, there has been more talk about subjects that appeal to the creative abilities within us. It was felt that the artistic subjects belonged to both categories, disciplined and recreational. At first, these recreational subjects implied resting after the academic work was completed; then they became an essential part of the work itself, as the activity was now judged from a different angle. It was recognized more and more that the artistic is an integrated part of the child's whole world.

The two schools of thought point to an essential development in the interpretation of what children need and what the task of the school is. Fortunately, in ordinary circumstances children have no need for recreation. All parents of healthy children learn this through many, often painful, experiences.

Children have, on the other hand, a need to be creative in everything they do in school. One thing, however, which may necessitate special creative lessons is a schedule devoid of creativity. This will cause fatigue.

Experience of such relationships must certainly underlay the quotation from the Danish "blue paper" which speaks of inner aspects in terms of "harmonious, happy and good people." One can question the validity of the adjectives in this quotation. Perhaps one would get to the essence of education by avoiding them and instead, combining the two quotations saying: "It is the task of the school, first and foremost, to encourage every opportunity for growing up, through activity and experience."

By formulating the educational guidelines in this way, one would have touched upon the central problem for all teaching in elementary schools.

Teaching at primary school age must be based mainly on activity and experience, the more so the younger the class. In adulthood, the sequence of the soul's expression is often that which Ibsen gives us in *Per Gynt*: "Think it, wish it, will it, too. But do it ..." As adults we live in our thoughts, which give impulse to our emotional life, which in turn activates the will.

For children it is just the opposite; they live out their daily lives in untiring playful activity. During the nursery school and primary school years, the child's emotional life becomes enriched and more complex. From that foundation arises the clear, awakened thought-life of adulthood. N.F.S. Grundtvig (1783–1872), the great Danish author and psalm-writer, put it this way: "What are thoughts but feelings that have become conscious of themselves?"

Or, in poetic form:
> ... since he has never lived
> who wise became from that
> which first he held not dear?

Activity and experience imply feeling and will, and when we consider that the life of thoughts is hidden within the expression "imparting basic elementary knowledge," we can see that the "blue paper" for Danish schools has actually incorporated Grundtvig by stating the three elements of the soul, which education of children is all about.

We go from activity of the will via the life of feelings to the life of thoughts, which ought to be the result of a long process. The teaching of children is thus a threefold process, with the development of concept coming last.

Many experiments with the learning process have been made in the past. Various direct approaches in the training of children's intellectual abilities have been attempted—for example through the use of teaching machines. However, it has been shown that physical activity is necessary if the child is to gain insight into the world of thought.

Many shortcuts have been attempted, and there are several biographies that show this. The life of English philosopher, John Stuart Mill (1806–1873), is one example. From a very tender age he was influenced directly by thinking elements. He made enormous progress in this area during the first years of his life, and he was considered to be an infant prodigy. However, it must be said that in his philosophy he shows an exceptional distrust of results obtained by thinking, and that he considered his non-existent childhood to have a direct connection with the difficult crisis he had to live through.

A child cannot grow by pure knowledge which is aimed strictly at the thought-life—not only because it is too difficult, but also because the child's development is via activity and experience. We may attempt to do this, and we will discover that some children, more than others, indeed are being educated. But this is because, when we stand before a class of children as teachers, we always plant a seed of activity, to one extent or the other, and this small germ goes through a transformation and becomes thought. This is the inevitable course of events for children.

It is of vital importance, however, that we as teachers not only launch the process, but that we remain aware of it throughout. Only in this way can we avoid wasting time in our teaching and become truly "effective"—perhaps not a good expression in this case, but it is an expression often used in the attack on creative and artistic ways of teaching.

Real effectiveness can be achieved only when we have a thorough knowledge of the soul's developmental process as mentioned above. One must know, in addition, that human development is not the sort of growth in which something is first small and then bigger, preserving the same basic structure. Rather it is one where abilities are completely transformed during growth from one level to another. A child is this kind of being, undergoing complete transformation on its way to adulthood.

In this regard the teacher must have great patience, not only from day to day when he must question the children on yesterday's lesson, but also from year to year, and even from one phase of life to the next. Probably one of the biggest obstacles a teacher has to overcome is to avoid wanting to harvest the next day that which was sown only the day before. Referring to this, humanitarian and educator, Johann Heinrich Pestalozzi (1746–1827), said:

Es ist eine große menschliche kraft, ohne ungeduld zu harren, zu warten, bis alles reifet.

It is a great human virtue to be able to wait without impatience, until everything ripens.

The strength needed to wait and while waiting to discover the exact rhythm which guides the child's growth, is the very strength which creates the proper "incubator heat" that surrounds all educational processes.

Jean Paul also talks of this "sowing" function in education, but he goes a step further and says that the educational task is more a question of warmth than of sowing. He has therefore concerned himself with the thought of Socrates, that we can only act as midwife for the abilities that are already latent in the child, awaiting opportunities for growth which we as teachers can create for them. That alone is our contribution as teachers, and this does not make the job any less responsible—on the contrary, only more so.

In arithmetic and mathematics the question then becomes the same as everywhere else: "What do the children bring with them?" and not, "What does society demand we put into them?" What laws of development must be followed if these inner qualities are to be brought out? Or put another way, it is not a question of creating something, but more of bringing something forth.

In our particular case we must ask, "What mathematics lies already buried within the child, and what are the rules for nurturing this already existing substance?"

We seldom consider what already exists in the child but are instead one-sidedly concerned with what we believe the child will later on become.

The child, being the starting point, demands a pedagogical study of us, whereas the adult student, the end result, requires a vocational, subject-oriented study—and it is the latter which tends to dominate our whole way of thinking.

We ought to occupy ourselves to a much greater degree with the laws of development and change which govern the interval between the starting point and the ending point, between the child and the youth. We try to avoid the idea of metamorphosis because it is difficult, and everything in our professional world supports us in this. It therefore requires of us a real effort of will to organize the teaching of mathematics in a way that it "arises out of the child."

Put clumsily, we must "avoid avoiding" the metamorphosis idea. This expression contains, however, something worthwhile—corresponding as it does in the world of mathematics to the well known "minus a minus equals a plus"—and suggests that we reach a higher level of positivity for the child through making such an effort of the will.

Chapter 3
NUMBER QUALITIES AND HIDDEN SECRETS
ESSENCE NUMBERS

We have spoken in the previous chapter about a child's sensitive periods and about relating our teaching to specific stages of development. A child's development is based upon the development of the senses, but the center of gravity moves quickly from one sense to another. If we use teaching materials that correspond to the rhythm of this development, we will be doing the whole human being an invaluable service.

Normally we do not speak of a "thought-sense" that senses thought in the same way as we sense sound, for example. However, if we do, we must refrain from using it until the child expresses a need for it, and not begin using it because we think the earlier, the better.

Neither do we speak of a mathematics or geometrical sense, but if we do anyway, it is not at all certain that it is connected to this thought sense, an impression we might get from many books on mathematics.

There is much in a child's life that points to the existence of such a mathematical sense, and that it functions from an early age. At the same time, it does not seem to be connected to the "thought-sense," if such a sense exists, for the mathematical sense does not experience the quantitative, with which we normally work, but much more the qualitative.

This we can understand by a careful study of the young child's physical expression or by studying the early history of mathematics.

We have plenty of evidence that people of earlier times considered mathematics to be a science about qualities. Traces of this can be found in our belief in lucky numbers and in the displeasure we show for certain other numbers. We meet this superstition frequently, and many of us shake the lottery ticket basket before we choose.

Thus we demonstrate that a series of numbers is not just a row of small, identical soldiers, each one a little bigger than its predecessor, but a row of individual figures linked together. The latter has more to do with our feelings.

For the ancient Greeks, too, the numbers had individual features. Number 6 was an especially perfect number. It could be divided by 1, 2, and 3, their sum being 6. It can also be divided by 6 according to modern view, but this was not counted—they regarded it as being within itself, so one could not divide by it. The number 6 thus had within it

$$1 + 2 + 3 = 6$$

and was therefore internally equal in value to that which it showed externally, its face value. There was thus perfection and harmony between inner and outer. 28 is the next perfect number, as its content is

$$1 + 2 + 4 + 7 + 14 = 28$$

If we try to find the next one, we will have to begin again when we reach 500, because we have gone a little too far. However, we have experienced in the process a little of the work that the great mathematicians had to carry out over the years in search of these perfect individuals. What effort, for example, must have been involved in finding the fifth number of this sort? The number is 33,550,366! A mathematician in Alexandria, Nicomachus from Gersa, sought it in vain, and allegedly lamented, "The good and the beautiful are few and far between, and easily counted, but the ugly and the evil are to be found in abundance."

What a different attitude towards numbers! There are beautiful and perfect numbers, and there are ugly and imperfect ones. 15 is an imperfect number; its inner content is only 9. It boasts with its outward face value. The content of 16 is 15, so it boasts only a little. 24 does not brag at all, as its content is

$$1 + 2 + 3 + 4 + 6 + 8 + 12 = 36$$

So, it is at once valuable and at the same time modest and unpretentious.

The number 360 plays an important role in many places. From the point of view above, it is a valuable number with many possibilities. Its content is 810 and one can express its value in the fraction

$$\frac{810}{360} = 2.25$$

while the number 24 has the value

$$\frac{36}{24} = 1.5$$

The perfect numbers all have the value of 1. So 360 becomes an important number, and not an insignificant personality. In comparison, 15 has the value

$$\frac{9}{15} = .06$$

Everywhere in the history of mathematics one meets these relationships. In geometry, for instance, there are relationships between the Platonic solids, which for the Greeks were connected with the cosmos and the four elements, and in these relationships one saw the two sexes mirrored. Generally, they experienced that the world of numbers and forms was closely related to the being of the human soul.

Young children are also deeply connected with such qualities, but because they are children, they are not able to describe their experience in words, nor are they able to relate these experiences to such difficult subjects as Platonic solids and perfect numbers. Their eye for measuring such things, however, is of the same kind. So they must function, so to speak, from the same corner of the world but express themselves in language corresponding to their age.

It is therefore of the greatest importance before teaching arithmetic that one attempts to form for himself a picture of earlier cultures. This is one way to prepare, in order to get ideas on how it might be possible to introduce children to the world of numbers. This would then be supplemented by looking for teaching materials that fit the skills of the children at their age level.

The Platonic solids played an exceedingly great role for the Greek mathematicians. It was not only Kepler who placed them concentrically within a sphere, but all throughout antiquity they were experienced in relation to the unity of the sphere. In the same spirit, we can let our children experience the somewhat simpler regular polygons inscribed within circles. If we keep to not more than 3 to 6 sides to begin with, then children, in the first round, can easily follow, and through this satisfy their sense for the qualities of numbers. We let them draw forms like these:

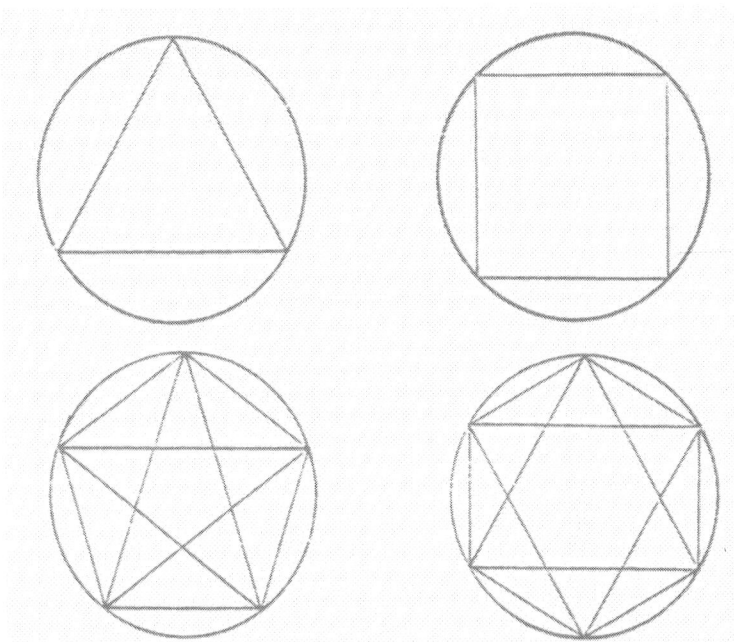

With five- and six-sided polygons, the children can draw the "stars" as well, which will eventually make it easier for them to achieve the regular shapes. Color, too, may be used in order to enhance the relationship of the numbers. Corresponding fields should be colored correspondingly; otherwise the picture will have a harlequin look about it, which is moving away from the original purpose.

The pupils in first grade know quite well that it is a question of making the sides equal if the figure is to be in harmony with itself—in the same way as the Greeks, who sought three dimensional solids that could be built up out of regular polygons. The endeavors are the same, the perceptions analogous.

It is important that children experience the unity, the oneness, the wholeness of the circle. Have them practice it first of all drawing freehand—the compass belongs to a much later stage. They can then practice the number 2 in this way:

When they have drawn a big, beautiful circle, they can be given nuts or chestnuts to place on the circle. We can give them 3 nuts, for example, and ask them to place them carefully on the circle. "Carefully" can mean many things, and all ways have to be accepted. However, the children should also be helped to place the nuts in such a way that the circle is divided into three 120-degree parts—without using this terminology with them.

In the same way 4, 5 and 6 nuts may be used.

When studying children in their efforts to divide the circle into 6 parts, for example, one can observe them in purely mathematical occupation. Here, work is being done which forms the basis for the mathematical thought processes of the adult. Let them remove 3 of the 6 nuts in different ways and experience the pure joy of seeing, in some cases, an earlier triangle suddenly reappear.

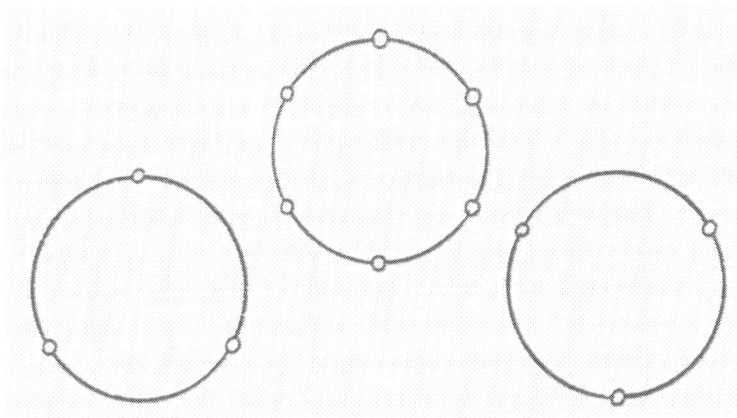

Six nuts give rise to two different triangles.

It may happen with the point upwards (for many, that is the nicest way) or with the point downwards. Some see the finest triangle with the point pointing upwards while others see, in the same triangle, the point pointing downward towards the left or downward towards the right.

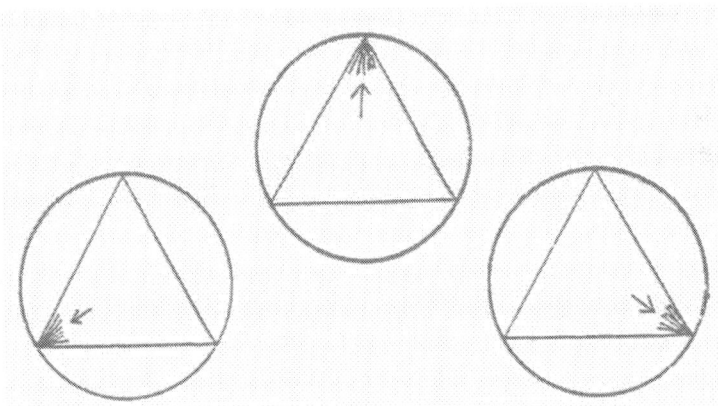

Triangle in motion

At last, one pupil sees that the triangle can move. His circle is just small enough, and his fingers long enough, for him to reach two of

25

the nuts with one thumb and middle finger, and the last nut with the index finger of the other hand. His shouts of joy summon the class to come see his triangle turn, so that the point is rotated from being above to being turned downward.

If you try this for yourself you will understand the excitement. During this process, one has to stand up in order to avoid twisting the hand. This creates some unrest in the class—when everybody wants to try it at the same time—but the teacher must get used to that.

Within the 120 degrees lies the quality of the triangle, appealing to our sense of harmony. We feel ourselves to be in balance when the three nuts are in their rightful places, and this experience is independent of the size of the circle.

The experience is quite different with the square, different again with the pentagon.

A hexagon is something very special for most children. Try giving them a handful of nuts after the first six have been placed and ask them to outline the two interlocking triangles. It is often a deep experience. If enough nuts have been given out, a figure will appear like the one shown in the drawing below, and it will give rise to great wonder.

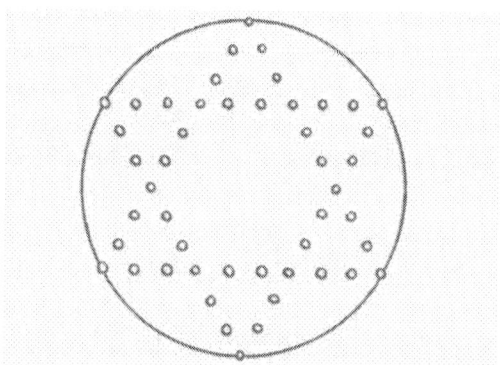

Six-pointed star made from nuts placed evenly.

Now there are lots of different-sized triangles, which, when the child aims to make them equal, create the hexagon shape in the middle. There is hectic activity with the hands and the eyes, and with the whole of the child's body, in both movement and balance. This lively activity develops the mathematician.

Let it be emphasized from the beginning: It is not a question of reviving old superstitions, but of teaching mathematics and arithmetic in the realm where it belongs, i.e., a long way from the functioning of the head. This will fulfill one of the child's greatest wishes, which is to remain in the world of bodily movement for as long as its laws of development demand.

Place 4 nuts in a circle! Now place an additional 4 in the spaces in between! Now draw a circle around the two squares with new nuts!

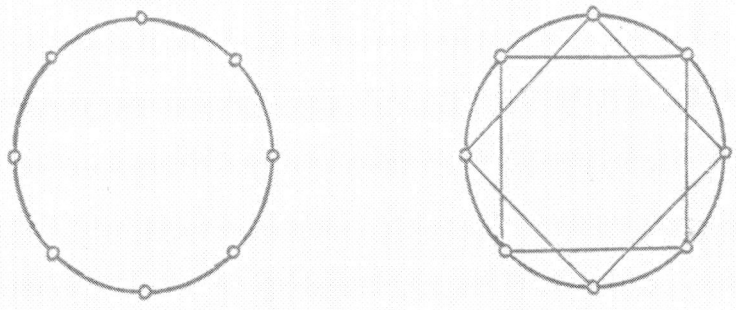

Eight nuts create two squares.

Now the triangles around the edge are quite different. This time they do not look like those in the hexagon. They are not quite so nice and the nuts are not so easily placed. It is not possible to place the nuts as regularly as before because the sides are not equal and cannot be measured against each other. The children's hands grapple with the irrational problem. They do not know anything about this and should not be told about it yet, but their eyes and fingers have discovered it. They may comment, "That's funny!", but the teacher will know that they will come back to this in 6 or 7 years' time.

Now make the square again. Then place two nuts evenly in each of the four spaces in between. It looks like a clock! Now remove every second nut. Look, our old hexagon is back again! If we then remove every second nut, the triangle comes back. From the clock we can derive both squares and triangles. This is an important discovery: 12 nuts in a circle contain many riddles!

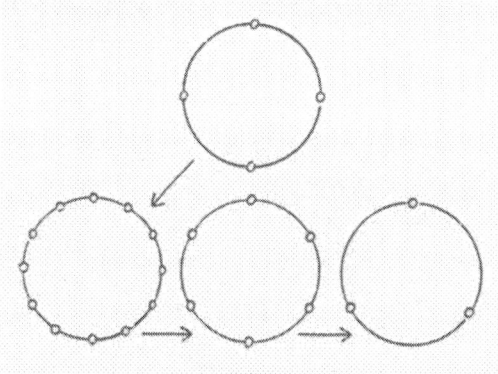

From 4 to 6 and 3 by way of 12

Next we place five nuts in a circle and then a nut in each space in between, which gives us a decagon. After looking at it for a bit, we decide to remove every second nut, twice in a row, and see what happens. The first time it works, but the second time the result is not so very good. We note this and have learned something about odd and even numbers.

If we place 16 nuts in a circle—which is not so difficult—we can then remove alternate nuts many times. Some children may say that 16 is a very even number, and in their own way have understood what "even" means. 20 is not nearly as even a number.

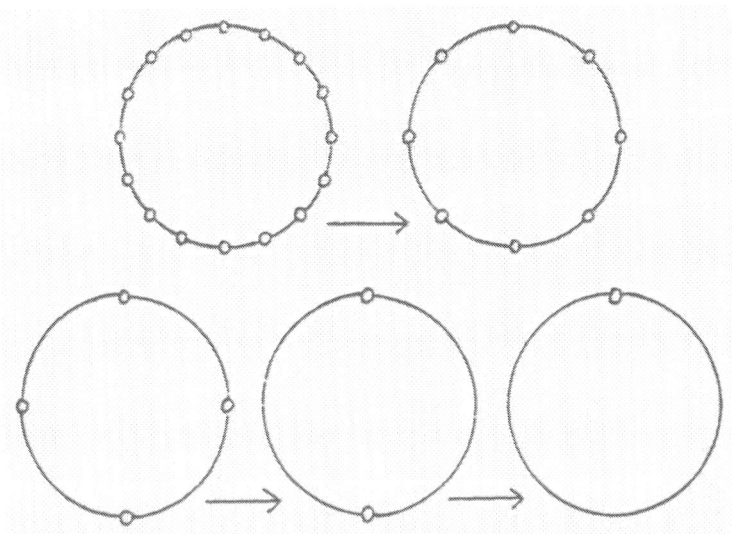

16 is an unusually even number.

Arriving at the next to the last figure, many pupils pose the question of whether we can go on any further: What is "every other" of two?

One day the children are given 7 nuts each, to be placed in a circle. They recognize at once some of the difficulties they had earlier when they placed out 5 nuts. They remember being helped by laying down on the floor with arms and legs outstretched which, together with their heads, made them into five-pointed stars. After that it was not so difficult to place the 5 nuts.

However, with 7 nuts the trouble begins because we are short a couple of arms. The head is easy to place. The feet must be a little closer together, and we must then imagine a double set of wings.

It is not easy to place 7 nuts in a circle.

At last our winged human being falls into place and we have a beautiful heptagon. If we now attempt to remove every second nut, we find again that it doesn't work. Instead, we content ourselves with tracing first with our fingers, then with a pencil, the path we would have taken if we had removed every second nut. Something exciting will then appear.

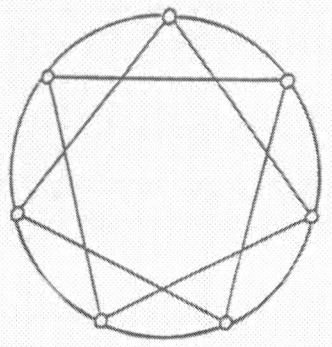

We connect every second nut.

And it is even more exciting, if each time we move to every third nut. This is difficult, but clearly worth the trouble.

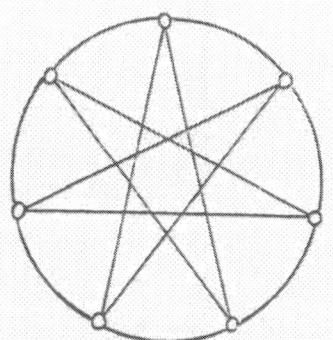

We connect every third nut.

By experimenting with the octagon in the same way, we can get a star made from two separate squares, but we can also get an eight-pointed star made from a continuous, unbroken line.

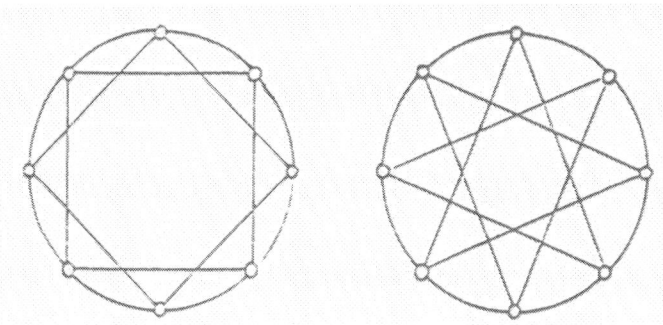

And it is the same with the nonagon, three triangles or a continuous line.

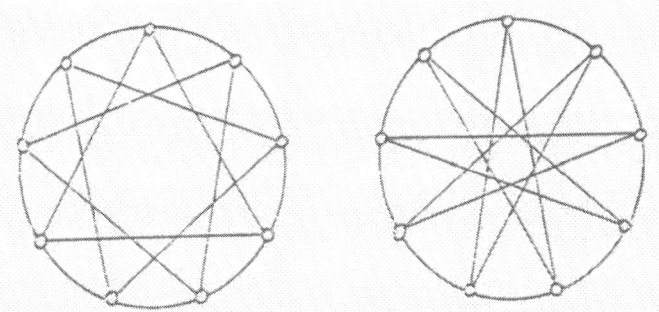

With the decagon, we have two five-pointed stars and a continuous line.

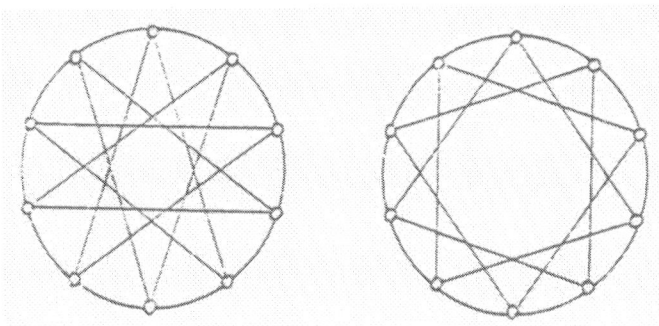

31

In this way, we experience that the five- and seven-pointed stars only arise from continuous lines. If we were to continue, we would find the same is true of stars with 11 and 13 points. Why is this the case?

The pupils will learn more about this later, but here they have already come into contact with the many questions associated with prime numbers.

The six-pointed star is also a special case, because it cannot arise from a continuous line, but must, of necessity, be created from two triangles, one that points upwards and one that is pointed downwards. It was not by accident that the Jewish people saw something special in this star. It has a deep symbolic meaning for them. However, we are concerned here only with the mathematical qualities that separate this star-shape from all the others.

These games with nuts should be used in parallel with exercises performed in a big hall, big enough for the children to make a circle holding hands. Or perhaps only half the class forms a circle, as it is also valuable to alternate between observing and being active.

Let us make a circle with 12 pupils. Now, if every other child takes a step forward or sits down, the picture will at once be different. If instead every third child takes a step forward, we will recognize our previous square. With every fourth, we get a triangle, and we see that all this works beautifully when there are 12 in the circle. With every fifth, we find that it does not work. 12 and 5 do not have much in common. When every sixth steps forward, something is created, but it is not as exciting.

We discover in this way that numbers belong together in special ways. If there are 13 in the circle, every third or fourth does not fit in, nor does every fifth. Every fifth does fit, however, when there are 15 in the circle, and what a beautiful triangle then appears! 15, 5 and 3 go very well together.

Returning to 12 pupils in the circle, an extra child goes around the circle unwinding a piece of string, of which every fourth child takes hold. At once, we have a beautiful triangle. In the same way we can easily create a square, by choosing every third child. 4 and 3, 3 and 4—easy to remember!

If we have two strings, we can make a six-pointed star from two triangles. First there is the one triangle, which we then lay on the floor while the second is being made. Then both are raised to the same height, and the star appears with unusual beauty. This then becomes an exercise in concentration, when we ask the children to rotate the six-pointed star around once. Complete silence will maintain in the hall while this takes place. The six pupils who are not holding the string can take a step back, thus marking the circle. The rest are making sure that the strings are tight, with everyone turning at the same pace, until they reach the starting point. The children who have marked the circle must also have a try, and now comes the most difficult: to pass the star over without spoiling it. This can probably be managed by a third grade class.

This is a serious game, and many things are learned. The figure contains six smaller triangles, around the circumference, which ideally must be equilateral. At the same time, they must be placed in a certain relationship to each other. The shape alters continuously during turning, and in order to correct any mistakes, the children must be aware of their neighbors' positions. Through this exercise, in making such observations and judgments, the pupils train in practical mathematical skills which are the real foundation for later mathematical ability at the level of thought.

It is difficult to ensure that the strings are constantly tight. This demands a great deal of cooperation and close observation of the movement of the other two pupils in the triangle. That in itself is very valuable. One of the strings may have an inclination to hang loose the whole time, and the triangle ceases to exist. In order to give the hands a feeling for what this means, a few days later the children can be given three long, wooden poles to work with. These replace the strings, and the children realize that the triangle has now become stable and cannot be altered at all. The same hands compare the two completely different experiences and "know," several years in advance, many things about triangle construction, about congruence and about why the carpenter often nails a diagonal beam across the rafters on a roof.

A diagonal beam prevents the rafters from collapsing.

These exercises for children in the younger classes, described above, can be supplemented ad infinitum. One may ask which part of the world of numbers they really illustrate. We distinguish, for example, between cardinal and ordinal numbers. The cardinal number 5 can be depicted with the help of 5 apples. The ordinal number 5 can be pointed out with the help of a single apple; it is the fifth in a line of apples, and we come to it by counting.

The fruit of knowledge, for better or for worse

The numbers that we have talked about up to now are neither cardinal nor ordinal numbers. Such characteristics can always be found within them, but they exist primarily on a different plane altogether.

We come close to this level for the number 5 when we say that 5 is a prime number. This characterizes it totally independently of whether it is used as a quantity or for ordering. Its neighbor, 6, is not a prime number, but a perfect number. Both have a special feeling about them, an individual identity, a feeling that has nothing to do with the fact that one of them is one bigger than the other. 5 is a prime number; this is part of its being.

This can also be experienced if one of the apples mentioned above is cut in half, not as if one wants to share it with somebody, but cutting through its equator. If you have never tried this before, and do so now, you will understand a little more of the quality of the number 5.

An apple is a fruit from the tree of knowledge, and it has been used throughout the ages to help understand both cardinal and ordinal numbers. A teacher will say, "Take 5 apples, for example …" and then explain about the two kinds of numbers. This is as it should be; it is a tradition. But they have always forgotten to cut the apples in two afterwards, to see what the true kernel of 5 is. Deepest inside it is neither cardinal nor ordinal, but an active power which of itself alone is able to create the two groups of numbers.

An apple cut through its equator – one hardly dares to eat it afterward!

We need to find a word that can describe this qualitative aspect of numbers. Together with cardinal and ordinal numbers, perhaps we ought to have "essence numbers" or "quality numbers" or "individuated numbers" or ... what shall we call them?

All early mathematics carries a message about this number, which is precisely why our children must experience, in one way or another, the quality in numbers—because deep down children have what we could call a historical need. This does not mean that there is a need for learning history—they have that, too—but the need to see with clear eyes those aspects of things in their surroundings that were important to previous cultural epochs.

The "blue paper" (see Introduction) says that the curriculum in the state schools is overfilled, and since that is undesirable, some of the material which is included purely out of tradition must be eliminated. There is considerable truth in this, but the reduction could easily be so severe that the child's energy sources may be damaged in the process. These sources need more consideration than is usually given. One becomes a little alarmed when reading that "only by a severe reduction of antiquated material" will it be possible to consider rapid growth in scientific research. The "blue paper" says that curriculum revision must be looked upon as an on-going process. Revisions may be undertaken at any time (without, however, disturbing the school's objective of bestowing confidence and harmony) by continually chasing the latest ideas.

It is undoubtedly true that there is a great deal of material that may be disregarded anyway because it has no connection with the inner child. However, the seeds of restlessness and nervousness are sown if the clearing-up is made on the basis of the child's having the same inner structure as an adult. The problems will be further aggravated if the process of metamorphosis, and the related need for tradition, are also ignored.

Children's immediate joy in fairy tales is intimately related to this and sometimes gives us the impression that our own children are more at home in the fairy tale world than in the technical world with which we so cleverly surround them.

In the very oldest myths, legends, and fairy tales, we can find real nourishment for the previously mentioned "historical" inner of the child. This inner makes the child in many ways more attached to previous cultural epochs than to us, in so far as we are merely members of a modern civilization.

Thus it is that we often find in fairy stories signs of earlier epochs' experience of "quality numbers," i.e., numbers which are neither cardinal nor ordinal, but creative powers within themselves—before they manifest as the one or the other kind of number. In fairy stories such "essence numbers" appear all the time, now in the shape of the two brothers, now as the twelve swans, now as the four winds or the three sisters. We might tend to lapse into number mystique, which we should not, but even in such a distorted form there is still a kernel of something timeless, a last, desperate holding onto deep layers within us.

In the following fairy tale from India, we experience the original number 4. It is about four brothers, and the question arises whether it is possible to imagine the four brothers being five, if a few more details were given. Or, does the number of brothers describe a relationship of things which, by its very nature, can be only fourfold and where the organic life of the subject simply dies if three or five is used? If the latter is true, then the number is an "essence number" in the sense described. Perhaps here we have found a new name for these numbers, namely organic or whole numbers. These names hint at the creative power behind the organic whole. One such number is the number four in this fairy tale of "The Lion."

Once upon a time, there were four Brahmin sons. The brothers loved each other with great devotion and decided to travel together to neighboring kingdoms in order to win fame and fortune. Of the four brothers, three had followed paths of deep, comprehensive study and had become experts in magic, astronomy, and alchemy—the most difficult of occult sciences. The fourth brother had learned nothing and had only his good sense.

As they wandered along one day, one of the learned brothers said, "Why should our ignorant brother take advantage of our learning? He cannot be anything but a burden upon our expedition. Never will he win favors from princes and kings, and he will therefore be a disgrace to us. Would it not be better to let him return home again?"

But the eldest brother answered, "No, let him share our good fortune. He is, after all, our dear brother, and we surely can find him a job which he will be able to manage without disgracing us."

So they continued on their journey together. Shortly afterwards they came into a forest, where they saw the old bones of a lion spread across the road. The bones were white as milk and hard as flint from being dried and bleached by the sun.

Now the brother who had earlier condemned the fourth brother for his ignorance said, "Let us show our brother what learning can do, by giving life to these bones and creating a new lion from them, so that he may be ashamed of his lack of knowledge. With a few magic words, I will gather together these dried bones and put each one in its rightful place." He then said the words, and the bones immediately drew together with a rattling sound, each to where it belonged, so that the skeleton was complete and whole.

"I," said the second brother, "can stretch the sinews across the bones, each one in its right place, with but a few words. And in between the sinews I shall fill in the muscles, colored red of the blood. And I shall create the vessels, fluids, glands and the marrow." Then he uttered the words, and there at their feet was the lion's body, complete, furry and huge.

"And I," said the third brother, "with but a single word can give warmth to the blood and set the heart to beat, so that the animal will begin to live, and to breathe, and to swallow other animals. And you will hear how he roars."

But before he had time to say the word, the fourth brother placed his hand over his brother's mouth. "No," he shouted. "Do not speak the word! This is a lion, and if you give it life, it will swallow us up."

But the others made fun of him and said, "Go home, you fool, what do you know of science?"

To this the fourth brother replied, "At least give your poor brother time to climb up into the tree before you give life to the lion." This they agreed to do.

He had just climbed the tree when the word was spoken, and the lion stirred and opened his big, yellow eyes. It stretched itself, stood up and roared. After that it turned toward the three wise men and swallowed them up whole.

When the lion had gone, the young man who knew nothing of science climbed down from the tree, unharmed, and returned safely to his home.

A fairy tale should never, of course, be interpreted for the children with abstracted meaning, but rather allowed to influence them in its own way. One must simply trust that the power it contains will be transformed together with the child and become its companion in the future. Therefore we shall mention only in passing that this fairy tale may be seen to point towards the fundamental human experience that everything in our surroundings can be divided up into four realms: an inert mineral world, a plant world having life but no freedom of movement, the animal world in which movement has been added as well as the ability to express oneself through sound and desire, and finally, the human world raised above the other three through the ability (for example), to foresee and plan. The animal—towards the end of the story—contains the three lower kingdoms and must therefore swallow the three brothers. The youngest can rise above not only their level, but also his own, and return to his original home.

The story illustrates a genuine foursome, not the 4 that is one bigger than 3 and one less than 5, but four with the structural power it received in the beginning of time from the Creator's hand.

Perhaps one might also speak of "structural" numbers!

The story of the lion and the four brothers could be told in many different situations, but certainly it would be appropriate in a first grade class. At the time of telling one should talk with the children about other things from their own lives which can help deepen their experience of the number 4. For example, the rhythmical course of the seasons can be brought to their consciousness. They might perhaps paint a landscape in which seasonal games are played, where the fields and the trees and the bushes are all in their seasonal clothing.

Then other things can be found which relate to the number 4, and where no other number can replace it. For example, we can talk to the children about why a book has four corners. First one must think through how it would be if books had three or five corners, or even more. Imagine how it would be to place them on a bookshelf, or in one's bookbag, or when they are to be wrapped and sent in the mail! Imagine also how comical it would be to see people in the library trying to find a certain book on the bookshelf if it were triangular or pentagonal!

No, books must have four corners. Four corners, or possibly eight—so that they will stand on a bookshelf and so that the title on the back can be read without breaking one's own back. But the quadrangular one is the very best—as any printer will quickly confirm. Imagine how difficult it would be to trim the pages of an octagonal book!

Well, let's put joking aside—but not without first seeing to it that this important aspect of daily life too is included, to the extent that it helps in experiencing number relationships, and to the extent understood by the age group. Humor, with warmth and sympathy, is always appropriate, even in mathematics.

Our own bodies provide a goldmine for experiencing numbers, especially in speaking with children. Younger children can still look at themselves and their bodies in a non-egoistic way. This changes when we come nearer to the age of 12 or 13. For the younger ones, however, pointing to number relationships in their own bodies is one of the very best ways of experiencing qualitative numbers.

We may speak to them about our four limbs, about the many ways in which they serve our body, and about the relationship they have with our surroundings. We can also tell them about our body's number 5 and its number 3. Later, we shall use the hand's number 5 and the two hands' number 10 as cardinal numbers, but do not stop with this! Let the children look at their hands as the beautiful sculptures they are. Let them experience the prayer position of the hands and appreciate the wholeness, the unity, that arises each time we greet each other by shaking hands.

Bear in mind, too, that perhaps later you will be teaching the same children the history of art, and then you will be speaking about the hands of Dürer, Michelangelo, and Leonardo. In such pictures one never speaks of a certain number of fingers but rather of an expression of the soul, which shows more clearly than anything else that a third kind of number, apart from cardinal and ordinal numbers, exists.

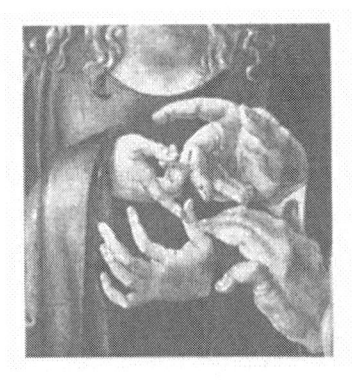

Study of Hands
Albrecht Dürer

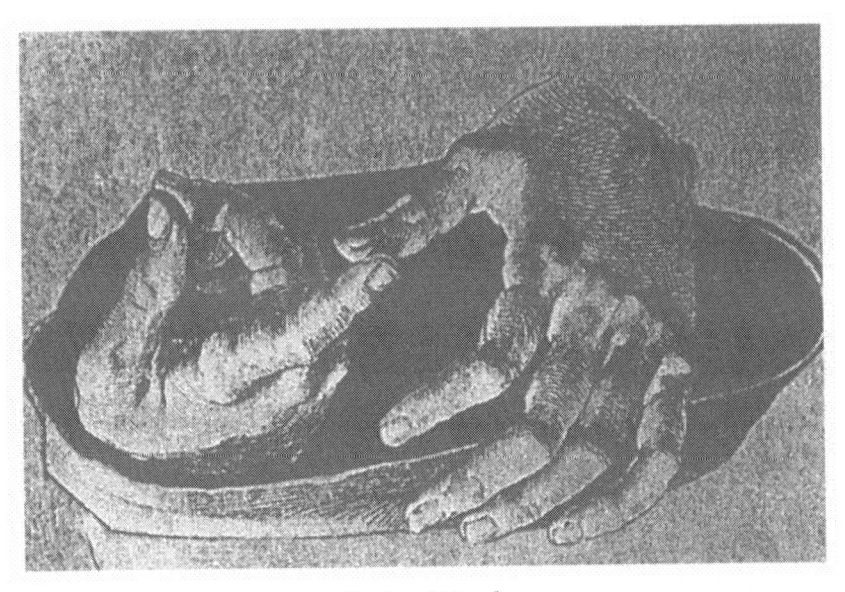

Study of Hands
Albrecht Dürer, 1506

Praying Hands
Albrecht Dürer

The number 3 is related to the Three Wise Men and also to the division of the body into limbs, trunk and head. There are also many "essence" numbers evident in animals, plants and minerals. For example, we find the number 6 in honeycomb cells, in monocotyledons, and in rock crystals.

The structure of a flower gives a wealth of inspiration. However, botanical relationships cannot be treated in detail in a first grade class. In fact, I must warn against bringing flowers and other parts of plants into the classroom in order to dissect them and count the numbers of petals or seed-pods. The experience of plants that should be conveyed to children at this age is best given by taking them for nature walks. In any case, let them begin in this way and then perhaps at a later stage continue in the classroom by recalling what they have seen. In this way their power of observation is being sharpened. It is important in the early phases of teaching arithmetic not to strive towards teaching cardinal numbers using the structure of a flower, but rather teach about structural numbers, about the creative powers at work in the flower as it is being formed and in the crystal as it slowly emerges. Teach about the powers which create these forms, which we adults count and classify in terms of numbers, but which exist in nature of themselves, as pure emanations, which existed long before any counting or human understanding came into being.

It is important to give the child experiences of this kind from nature, before the experience of counting. And then, to remember Pestalozzi's advice, quoted earlier (in Chapter 2):

"It is a great human virtue to be able to wait without impatience."

Chapter 4
RHYTHM AND NUMBERS

Until now we have talked mainly about exercises based upon form and have been, therefore, in the dimension of space. Now we go into the dimension of time and add new exercises where the emphasis is on beat and rhythm.

In each class it is important to cover both aspects, as any sizable group of pupils will always contain some children who are dominated by visual and some by aural abilities. The main thing, though, is that these rhythmical games lay the foundation for the multiplication tables and many other relationships in mathematics. And while the previous group of exercises had a close connection to cardinal numbers, these rhythmical exercises are closely related to ordinal numbers.

Last but not least, these games satisfy one of the deepest needs of children, the need for experiencing rhythm in their surroundings, an element which is sadly neglected in our time. There used to be a natural division of the day, for example, work time, meal time, play time, reading time, quiet time, and bedtime. All these have become vague and uncertain, and the result is that a tremendous insecurity has developed, in the wake of which restlessness and nervousness may be experienced. It is therefore important to find new areas for the expression of rhythm, and in connection with arithmetic, it ought to be an entirely natural matter to cultivate this sphere, since arithmetic and rhythm have a very close, but almost completely overlooked, relationship to one another.

Mathematics and music belong together. The laws of acoustics show this very clearly. It is also well known that many of the great mathematicians were musicians, and that they felt their musical work inspired their thinking activity. In music one experiences mathematics unconsciously. Gottfried Leibniz (1646–1716), German mathematician and philosopher, expressed it in this way:

> Music is a mathematical exercise of the soul, in which the soul is unaware that it is dealing with numbers. The soul gains knowledge in many unclear, unnoticed perceptive activities, of which the average attentive observation cannot be aware. Those who believe that nothing can take place in the soul without they themselves being conscious of it are mistaken. Even though the soul itself is not aware that it is calculating, it nevertheless feels the effects which these unnoticed calculations exert, whether it be the joy experienced through harmony or the restlessness of discord…

In school it is always exciting when teaching acoustics, for example in the sixth year, to discover that it is possible to find the middle of a string with the aid of the ear alone. When the ear recognizes the octave from the tone, the finger has found the middle. A well-developed ear is very discriminating, and it is also easy to find two-thirds and three-fourths by listening to known intervals.

Music, however, is not only a matter of tones, but also of rhythm and beat. It is here that young children are just waiting to be activated—they love to move their bodies and limbs.

As we said earlier, arithmetic, is a subject which the adult associates with intellectual dexterity. It has always been necessary, therefore, to work very hard preparing it to make it acceptable to children. This situation characterizes us more than the children—namely, our lack of understanding of children and of the stages of metamorphosis. Had we such an understanding, we would begin teaching arithmetic in the school gymnasium, because arithmetic begins with the body, arms, and legs.

Arithmetic begins in chaos. All the old mythologies tell us that the world also began as chaos, and that little by little the cosmos slowly emerged. This is also the way of arithmetic and mathematics. And the deeper the layers within the human being we are able to activate, the more well-founded its cosmos will become.

Leibniz knew this. He was fully conscious that deep down in the soul, a process takes place of tremendous importance for the whole individual—and that we only recognize a very little of what we, as whole beings, really are.

Leibniz was one of mankind's greatest mathematicians. Had fate wanted it differently, he might instead have become one of the best swimming instructors! For in both areas it is of the utmost importance to be fully aware that the movement visible on the surface is not the whole story, that underneath are movements of great consequence to us—now filling us with joy, now with sorrow—despite their being hidden from our view or difficult to observe.

Children first experience mathematics in this way, under the surface, unconsciously, in "unclear, unnoticed, perceptive activity." Our task is to find a suitable classroom for these first experiences in mathematics, a suitable "swimming pool," so to speak. Why not the school gymnasium, for example?

Having arrived there, we join hands and once again make a big circle. We move around while we slowly learn not to talk but listen instead to the sound of feet against the floor.

By listening to the feet in this way, the children will gradually move to the same beat. We learn to make the feet speak more loudly or more slowly, sometimes changing abruptly and sometimes gradually. It does not take long before we can follow the teacher's hand signals. Sometimes he raises his hand, sometimes lowers it, the feet stamp, and then they walk softly across the floor. We learn the sound of the heels and that of the toes.

But we have more than just heels and toes—we also have a left and a right foot. And they move not only forwards and backwards, but also to the side! It is important to be conscious of this now, as we shall need it later on.

Now we practice using left and right. We stamp strongly with the right foot and tread lightly with the left. If we move clockwise we will be stamping strongly towards the middle of the circle and lightly towards the outside. The "music" which arises is very different from the previous one, and by listening to both we realize that we now have two melodies.

We can also play the same melody by marking time, in place and using the feet as before. We can begin by moving forward to the melody and then, on a certain signal, walk in place, attempting to keep the sound the same. In this way we learn to listen. We then let a pupil stand outside the circle with his back turned and ask him to hear the difference, so that we must take even more care to make it perfect.

49

Next, we all turn towards the middle, while we walk in place to this heavy/light, heavy/light rhythm. This is more difficult. This time there is no help from the person in front of us. On the contrary, it looks as if she or he is doing just the opposite, although she or he is actually doing the same. It is not easy when facing one another in a circle, but it is the same sort of "crisscross" as when we shake hands every day, both of us using our right hands.

Some confusion can easily set in. It is also intriguing to glance around the circle and see the movement slowly change from one person to the next, from the person just opposite doing the opposite, round to one's neighbor doing exactly the same as oneself.

Finally we come to the most difficult movement. We use the feet in the same way as before, but now one person at a time: light/heavy by the first child, then by his neighbor, and so on around the circle. It is not easy for a child to stand still while the others are doing it, and so, when it comes to his turn, the stamping is often too quick, and it sounds all wrong. But slowly everybody gets it right, and in the end one need not even think about it. The melody itself tells one's feet what to do.

The first counting exercises can be introduced to the children during their first game, the one in which they listen for the beat. There are always many children who can count even before they begin school, and the less skilled will quickly learn in this exercise. We start by counting to ten, taking a step for each number. On ten we stop and jump with our feet together. It is the same game I played as a child, where we counted 10 - 20 - 30 - 40... until on 100 we fell backwards into the waves, overcoming our fear of the cold sea.

Here, we land on both feet on ten, looking at each other proudly. We can also stop on eight, but this might be more difficult, because we need to know in advance when it is coming—and eight can sometimes come much sooner than one thinks!

Five comes even quicker, but when once we have learned it well, we can also try it backwards, saying 5 - 4 - 3 - 2 - 1. This is very difficult at first! The legs seem almost frozen, and it is as if each foot hesitates before coming down on the floor. Finally we all get it, but then some may take one extra step and say "zero." No matter—we can all try it, and jump on "zero" with our feet together, as before.

Now! 1 2 3 4 5

Eventually we can all do it. The teacher says "now," and the children go into action:

Now! 1 - 2 - 3 - 4 - 5   5 - 4 - 3 - 2 - 1 - zero.

We will return to the little words "zero" and "now" later on.

Now! - 1 - 2 - 3 - 4 - 5
Zero! - 1 - 2 - 3 - 4 - 5   pause

Having learned to count to ten, the next step is the more difficult stretch up to twenty. But, again, if we let them familiarize themselves with the new words while walking and with number rhymes, the children in a first grade class will soon be able to count to twenty with certainty. In

English and German it is easier than in Danish to learn the numbers from ten to twenty, and if we simultaneously use the language lesson to let the children taste the sound of the word parts, together with the introduction of spelling them, they will soon make the connection between twelve and two, thirteen and three, fourteen and four, etc. Some of the words may still be difficult to understand, but with the help of the feet and the rhymes, they too will come.

The children now feel proud and happy to be able to say so many numbers, one after the other, without being interrupted. This is one of the peculiar things about counting out a series of numbers. Much of what a seven-year-old says may be right according to some adults, but wrong according to others. The very same words! Some grownups shake their heads and interrupt, others smile and nod encouragingly—to the same story. Counting, however, is different. People don't have opinions about it. Once learned correctly, everybody nods approval. They may tire of hearing it repeated as often as we do in school, but everybody likes to hear it once. And if they do interrupt, it is not because we have said something wrong, but for some other reason.

Counting sequences are one of the fundamental experiences in a child's life, and there is hardly anyone who has not at some point experienced the pure joy of repeating one aloud. This is because a child finds satisfaction in recounting stories and experiences. Counting and re-counting have a common basis. We all know about the curiosity that leads children to go on asking questions into the absurd, so that eventually we have to stop them because we are tired, and they could go on forever. The child himself may even be unhappy but does not know how to stop and get out of his own game. With numbers the situation is different, because here one does not end up in the absurd. It is marvelous to know that if one has the necessary energy, it is possible to continue indefinitely without moving away from reality. Each time we say a number, we already know in advance that another is waiting, just after. We even know what it is called. If we do make a mistake, it is not only the grumpy few who correct us, but also those who are friendly. The correction is always exactly the same, there is no disagreement. In arithmetic there are no arguments. If one doesn't understand a particular thing now, it can be understood later—if ones takes the trouble, of course. One's own understanding corresponds to that agreed upon by everyone. An agreement which is not reached by voting, but by hard work.

In other countries people may call the numbers by different names and what they say may sound different, but that does not mean they disagree with us. Behind both their words and ours lie the same numbers in reality—our different expressions mean the same thing. Reality is thus a strange thing, something in which we are involved in one way, yet completely apart from us in another way.

There lives in each child an intuition of this, and behind the spontaneous joy of reciting numbers lies the seed of security through knowledge of a reality upon which we all agree. A reality which is not forced upon us but is accepted because we know that here begins the only freedom which has any hope of lasting, because it does not violate another's freedom and never creates disagreement.

Therefore, let the children count, and count and count. Here they have their first opportunity for conscious contact with the infinite, and this experience is one of the most important steps in acquiring a living, useful, appreciation of mathematics.

In connection with counting, the children must become acquainted with the decimal system, and this can be done in an infinite number of ways. If we take the starting point to be their joy in the sound of language, we might use the following "number-runner" exercise.

"Number-runner" denotes two things. First, it is a line on the floor with the individual numbers marked off and extra large marks for the multiples of ten—perhaps in a different color.

A "number-runner" is also the pupil whom we let run or walk along the number line while saying the corresponding numbers out loud. He counts to ten while stepping off the corresponding steps. On "ten" he makes a brief stop and, just as he continues on, another pupil starts out from zero, both counting aloud simultaneously. Everybody can hear the difference in the two speakers, but also the prevailing similarities. Without a lot of explanation on the structure of the decimal system, the rhythmic nature of the system is learned in a way which reaches into deeper levels in the pupils than any theorizing could possibly do.

Things become even clearer if two children start at the same time, one from zero and one from twenty. Or let four children start simultaneously from zero, twenty, thirty, and forty. Through this the children will experience the structure of the system, especially if one can get the four pupils to count in unison. Finally, a fifth child can be added, starting from number ten, through which the special nature of the interval from ten to twenty may be experienced.

Later in the first year, we can place pupils on 10 - 20 - 30 - 40, etc., and ask a number-runner to start out from the beginning: 1 - 2 - 3 - 4, etc. When he arrives at 10, he takes the hand of his waiting friend who says "ten," to which the number-runner replies "and zero." Then they move on together, and on 11 the friend says "ten" again and the number-runner says "and one." On 12 the sounds are "ten and two," then "ten and three," "ten and four," etc. To the children this sounds a little odd, but they know it must be right! Arriving at 20 in this way, the number 10 delivers the number-runner to number 20 who says "twenty," whereupon the number-runner says "and zero." Those two then continue on together, while number 10 returns to his original starting point to see if any more number-runners turn up. This does not happen, and so he observes what is happening to 20. Here the number-runner has taken a step forward, and the sounds "and one" are heard. This time it is not only correct, but it also sounds better—it is almost the way we normally say numbers. This will continue as we work our way towards 100.

Returning to the interval between 10 and 20, we now know that 11 means 1 and 10, 12 means 2 and 10, etc. It is now possible to hear the real numbers. In 16, for example, one can hear the number 6, and the "teen" which follows must mean 10. This is easy to hear afterwards!

Understanding comes afterwards, which is how it should be. First, there should be activity, then comprehension. First knowing how, then knowing why.

Later, the position system must be clarified for the pupils, and it will then be a great help for the teacher if he can recall, together with the children, their having played the number-runner game. They will then remember that from number 10 onwards, the number-runner was never alone, corresponding to the fact that above ten we write more than one numeral.

At a later date we can consider with the children how the game might have gone on if we had continued past 100. It is probably best to modify the game by letting the pupil at 10 continue right through to 100, rather than having pupils at all the intervals of 10. At 100 he would meet a third person who would then continue. If we imagine how this might look at the start of the exercise, it might be like this:

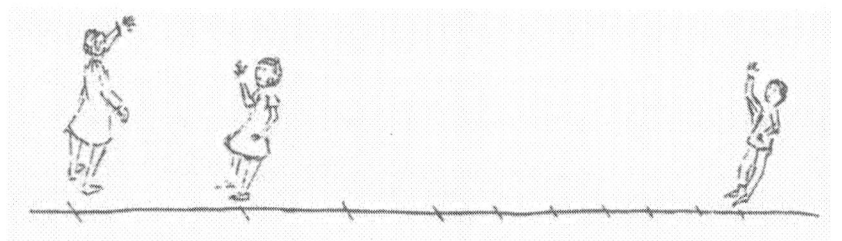

Three children are ready to practice the decimal system.

It is possible to illustrate many phenomena of numbers starting from this simple game. Here we have only tried to show that from a certain point on, the number-runner was never alone. Because of his position, he was necessarily "tied" to another person. This connection must be emphasized. It can be done, for example, by letting the children link arms or by using a red ribbon to "tie" them together. They can only be released when they pass backwards over 10.*

Another game with the same aim is the following:

On the floor we draw a number-runner up to 100, divided into 10s. The first pupil runs with very small steps while counting up to 10. At 10 he momentarily meets another runner, who meanwhile has moved in a long arc from 0 to 10, saying only one word: a long drawn out "oooooone." His "ne" falls together with the first runner's "ten," and he thus represents the 10s. While the unit-runner hurries on towards 20 with quick numbers, the 10s-friend has started on a second long curve, saying in his own slow way

---

* In Danish or German or other languages where the 20, 30, 40, etc. numbers are spoken in the reverse way, e.g., one and twenty, two and twenty, three and twenty, the exercise should be changed appropriately, letting the number-runner speak first, "One" followed by the 20-friend's "and twenty." then covered one of his own units, while the second, moving on the tens, has been through 10 of his units, and the third has covered 100 small steps.

"twoooo." Thus it continues. The third runner proceeds in an even bigger curve saying very phlegmatically "ouououoou-u-u-ne," because he must reach his last "ne" only when he meets his two companions on 100. He has then covered one of his own units, while the second, moving on the tens, has been through 10 of his units, and the third has covered 100 small steps.

The fingers correspond well to the decimal system – or the other way round.

Now this pattern must be transferred to writing numbers with the units, tens, etc., in the right places. One way of doing this is to draw three number-lines side by side on the floor and place "sleeping" pupils on number 1 of the first line, on 10 of the second, and on 100 of the third line. If one now awakens the pupil on 1 and lets him walk his line and awaken the pupil on 10, together they can move in parallel to awaken 100 when they get there. The distance must be kept more or less proportionally correct. While walking each should have a pad with big numbers on it, which they flip over as they go. This will mean that everybody can also see the numbers as they are reached. The question of which figure goes on the

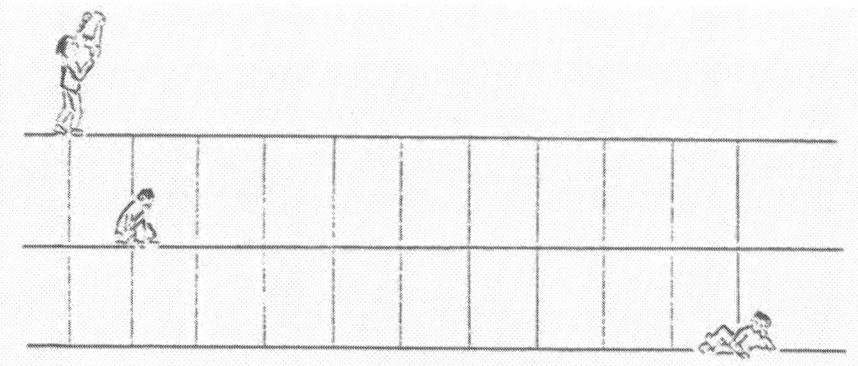

The position system with three number-lines

right and which on the left does not cause any trouble, because everybody saw that Janet with her units stood to the right, and John, with the tens, was to the left. At the same time, the children can see how busy Janet is with her pad and that John has plenty of time. And whether it was Janet or John who was busy, that is a thing which is easily remembered. Had the movements existed merely in thought, then only a few children would remember them the next day.

Another method of illustrating the position system is with only one number line, with the children walking sideways from right to left, facing the audience with their pads in front of them. They proceed in the same way as before, awakening their friends when they reach them and pushing them along with their right shoulders. In this way the positions are retained.

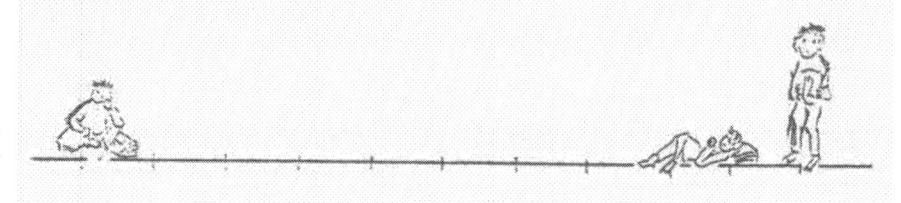

The position system with one number-line

Now let the children form a circle, facing the center, and then count around the circle, each one saying his number. The children already know how many pupils there are in the class, if all are present, and therefore know when they reach the end. If we ask individual pupils what their numbers are we will discover that only a few can remember. In a way they have placed themselves quite unselfishly as part of the recital as a whole and are unaware of their individual positions. If we try the sequence again, everybody will be able to remember his or her number. However, this time the rhythm of the counting is lost because of each individual's awareness of his or her own number.

Now it is possible to question the pupils which number they have, for example by pointing, in such a way that certain numerical relationships become apparent. For example, we can hear the numbers 5-7-9-11-13, etc., and having been around the circle once, the pupils can continue this series of numbers onward as if there had been more pupils in the class. Starting instead with 2-4-6 in this way, we can allow the pupils to hear a very definite 2 times table, even though it has not yet been introduced. We can hear the beginning of the series 1-3-2-4-3-5 by pointing in turn at the appropriate pupil, and could just as well have said, two forward, one back. Soon the series can be continued without pointing, and the pupils feel they are very well oriented in the world of numbers, with both their eyes and their ears.

In another exercise we let ten pupils make a circle, while the rest of the class watches. This time we count to 30. The pupils have been told in advance that they will have several numbers to remember, since we will go around the circle more than once. This improves with practice, and everybody will be able to say his or her number when the trick has been discovered.

For example, which numbers did Peter have? Well, Peter has a good memory, and he had 3, then 13 and finally 23.

Which numbers did Joan have? She had 9, 19 and 29.

And John? Well, John has a poor memory, and his attention is inclined to wander, but otherwise he is a quick lad, so he soon realizes that he must have had 7, 17 and 27.

When they try it again, it becomes obvious that John does indeed have 7, 17 and 27. He suddenly becomes very attentive, even if it does take a long time to go around the circle three times. He has never shown so much concentration before!

The next day the exercise is tried again with the same pupils, and everybody knows what will happen. However, the teacher then joins the circle, because he also want to have a go! He places himself between 7 and 8. John knows at once that something will have to change. Before the second round he knows for certain what will happen, so that he spoils the rhythm completely, when it is his turn to say his number. It is 18 this time, and a moment later he whispers to his neighbor (on the left) that the next will be 29.

When the teacher has listened to numbers from other pupils (1-12-23, 5-16-27, etc.), he quickly moves in between 6 and 7, and the game begins again! Nobody realizes he has moved, except John and himself, and he hurries to finish the game in order to calm John down!

Games of pure counting, and the consequences that follow from counting in certain ways, should play a big role during the first years of school. This is still true even after the pupils feel secure and sure of themselves and how the number system works. The best games are those that begin from standing or moving in a circle. They have a timeless character, because one is standing in a line with no beginning and no end.

And there are many possibilities for transforming the circle. Both of these aspects are important in mathematics.

However, there are also many possibilities if we place ourselves in a long row in the hall. For example, 10 pupils stand in a line, one behind the other, and count to 10 so that each pupil has a number. When the last pupil says "ten," he runs to the front of the line and says "one," wherefore the other pupils continue to 10 again. The one who is now last runs up to the front and says "one" when he is in place. The game continues in this way until the first pupil has reached the last place at the back and says 10. This generally does not pose any difficulties. Later on the children experience that what the row did many times, namely count from 1 to 10, was also done by each individual, only once and much more slowly. (They can be told later that it was ten times as slow.) However, it was not the same for everybody as no two children said it in exactly the same order.

The first one had said slowly: 1 - 2 - 3 - 4 - 5 - 6 - 7 - 8 - 9 - 10
the second: 2 - 3 - 4 - 5 - 6 - 7 - 8 - 9 - 10 - 1
the third: 3 - 4 - 5 - 6 - 7 - 8 - 9 - 10 - 1 - 2
the fourth: 4 - 5 - 6 - 7 - 8 - 9 - 10 - 1 - 2 - 3, etc.
and the tenth: 10 - 1 - 2 - 3 - 4 - 5 - 6 - 7 - 8 - 9

At some time or other the children should write down these series of numbers from beginning to end. They contain many regularities, and children enjoy searching for the rules and making new discoveries in such tables.

The following day, one can let the pupils say aloud and together the series of numbers for the seventh pupil, i.e.,

7 - 8 - 9 - 10 - 1 - 2 - 3 - 4 - 5 - 6

On "six" they will close their mouths tightly, to show that they know very well that number 7 must not be repeated. In the silence which follows the teacher quickly asks the children to imagine that there are 12 pupils in the line. What numbers would the fourth child then have to say?
The children answer aloud, together:

4 - 5 - 6 - 7 - 8 - 9 - 10 - 11 - 12 - 1 - 2 - 3

There is another game that can be played, perhaps when the children seem to be tired. When asked if they really are tired, maybe three will say no, and so they are called up to the front of the class. If the class is asked how far it is possible to count with three pupils, they will probably answer "three." However, if the first pupil gives his number "one," he can then run to the back of the row (especially if the counting is reasonably slow), so that he is ready to say "four," after "two" and "three" are said. Meanwhile, the second pupil runs to the back ready to say "five," and so on.

Everybody pays great attention when the game starts. The tired have forgotten their tiredness and, in great excitement, follow the counting to see how far it is possible to go. To avoid confusion, it is a good idea to clap on each number, ensuring that there is a firm, slow rhythm, so that there will be time to say longer numbers when we have passed 20. Eventually we all get tired and stop.

Another day, when we are tired from running around again, we will hold a number-conversation in the classroom, between the teacher and the pupils.

The teacher (T) gives a number or a series of numbers, and each pupil (P) continues the series, using the same number of numbers.

T :	1 - 2 - 3
P :	4 - 5 - 6
T :	7 - 8
P :	9 - 10
T :	11 - 12 - 13 - 14
P :	15 - 16 - 17 - 18
T :	19
P :	20, etc.

This can be done with differing intonations, speeds and volumes. In this way it becomes a dramatic conversation which the children enjoy. Here one may be worried, short-tempered, angry, lazy, or even indolent. One is also allowed to mock-scold the teacher, returning his intonation with vigor!

One can also take the same number of numbers every time. This makes it easier, and at the same time we are working with the multiplication tables without really knowing it.

T :	1 - 2 - 3
P :	4 - 5 - 6
T :	7 - 8 - 9
P :	10 - 11 - 12, etc.

However, we have as yet not talked about the tables, and when we do begin with them, it should not be in the classroom, but rather in the school gymnasium or another large, empty room.

Chapter 5
MULTIPLICATION TABLES

It is often said that multiplication tables are dry and abstract, but at the same time they can give a great deal of pleasure and activity.

Knowing the times tables is certainly a basic necessity to various parts of adult daily life, and because we have to a great degree forgotten what children really need and project our own needs onto them, we teach tables as if we were trying to make bread without yeast! This bread then becomes a small, dried-up lump, and it takes a great deal of saliva and swallowing to get it down.

Let us, therefore, begin instead with movement and rhythm, with the music of mathematics. The children stand, once again, in a circle and go round clockwise. On every third step they jump with their feet together. Left-right-hop, left-right-hop. With the children's past training, they soon have a fine three/four rhythm, and when the beat is firmly established, they might perhaps begin to sing:

> Oh, my darling, oh, my darling,
> Oh, my darling, Clementine,
> You are lost and gone forever,
> Oh, my darling, Clementine!

or some other well known song in 3:4 time. The problem here, mostly for the teacher, is to get the song started right so that the hop comes on the right syllable, in this case on the "dar" of "darling."

Now we can try something that is more difficult, but with which the body feels more at home. The children now go left-right-hop, right-left-hop, so that the legs alternate. Each rhythm has its own quality. As children we experience that quality, before we are consciously aware of numbers. We experience this particularly when we change to a four/four rhythm; here is a completely new world in which to develop, something which every composer knows. The two rhythms are totally different musical expressions.

This time we will go left-right-left-hop, left-right-left-hop, etc. We can also alternate the left and right feet after the hop, but this does not follow as easily, or as automatically, as in the three/four rhythm—it even feels somehow wrong.

The class will need a lot of practice with these two rhythms before they move on to the actual tables. First they must really experience the "music," the different feeling of the rhythms. And later, from the rhythms, let the numbers come naturally. We can do this quite simply by singing numbers instead of the words to the rhythms ... e.g.,

1-2-3 - 4-5-6 - 7-8-9, etc.

with a hop on   3 - 6 - 9, etc.

With the help of the teacher it will therefore be:

"Now"   1 - 2 - 3 - 4 - 5 - 6, etc.

That little "now" is very important, and we will return to it later.

If we continue to 30, the numbers become longer and more of a mouthful. We then decide to say them aloud only when we hop, thinking the numbers silently in between:

(1) (2) 3 (4) (5) 6 (7) (8) 9, etc.

The spoken numbers are 3 - 6 - 9 - 12 - 15, etc.

Our feet on the floor, however, sound all the time—even if our mouths are occasionally silent. Our feet make good connections between the numbers. They help us to understand what the tables are all about.

It has been said of music that the greatest importance is to be found less in the actual notes than in the effort our souls must make to move from one tone to another. The notes are like the milestones we find on the roadside, but it is the space in between them which moves us along. In fact, it is the inaudible, within our own activity, which gives the true content to what we are doing.

It is the same with the tables. It is not the numbers we experience, but the space to be found between them. It is this which gives each table its own particular character, and our feet help us to understand this.

If we merely teach our pupils the numbers 3 - 6 - 9, etc., it would be the same as teaching them the mile markers between Boston and Washington, which would not be much of a geography lesson!

Our thoughts can jump over the spaces, but our feet cannot. On the contrary, our feet are a yardstick for the intervals, and it is here that the children live.

The whole problem lies in bringing the children to experience these intervals. This is the correct way to learn, later on, where the borders lie, that is, the numbers themselves.

Once again we return to the four/four rhythm, this time saying the numbers and hopping on every fourth number:

1 - 2 - 3 - **4** - 5 - 6 - 7 - **8** - 9 - 10 - 11 - **12**, etc.

Later we will say aloud only 4 - 8 - 12, etc., but still walk to keep the intervals in between alive. It will soon be apparent that the intervals between spoken words are longer and so demand a greater degree of concentration than before to maintain the rhythm. However, it is still quite easy, and we begin to feel that our legs do it almost on their own.

Going from 4 to 5 we encounter difficulties. We experiment walking and speaking aloud:

1 - 2 - 3 - 4 - **5** - 6 - 7 - 8 - 9 - **10**, etc.

When you do this yourself, try to forget all knowledge of the 5 times table. As adults, we know that we are going to hop on 5 - 10 - 15 - 20. The children, however, do not know this sequence, so they learn a rhythm, and this one is not easy. It is worth remembering that for only a very short period in a child's life does he or she have the chance to experience this rhythm as a pure rhythm, because the 5 times table is so easy to learn.

The 6 times table is easier, because it can be divided into 3 plus 3, but this is not something we need explain to the children. They simply experience that it is easier or, perhaps, that they are becoming more capable. The six-rhythm can be divided into two with the same ease with which we divide a six-pointed star, made up of two separate triangles that cannot be drawn with one sweep of a pencil. Here the cooperation between two and three becomes apparent. A tripartite rhythm is easy to master. We can manage two of them, one after the other, as easily as we can see how the two triangles fit into each other. For the first time, the children experience cooperation between two numbers, and for a brief moment they have a glimpse of a whole new world.

This reminds us that we have skipped over the 2s table, but it can be easily learned in the circle again:

Left-hop-left-hop, etc.

or better still: left-hop-right-hop-left-hop, etc.

and then: 1 - **2** - 3 - **4** - 5 - **6**... etc.

ending with: (1) 2 (3) 4 (5) 6, etc.

We now break our circle for the first time. A number of the children stand at one end of the room, holding hands. They move together down the room saying their 2s table:

(1) - 2 - (3) - 4 - (5) - 6, etc.

Before 20 they are certain to have reached the opposite wall, but since we want to learn through 20, we will need to go back to the beginning and try again, using smaller steps. Eventually we learn to use the correct sized steps, so that we reach the middle of the room on 10 and the opposite wall on 20.

Now we can try the same exercise with the threes table. This is even more difficult—perhaps it is a good idea to go outside where no other class can see us!

Next we come to the 4s table, then the 5s table and at last to the 6s table. This last table is very difficult to walk in this way. On the other hand, the silent counting can be done by moving the hands together, and the children enjoy seeing the whole line jumping forward simultaneously. It will be noticed that while silently counting, there is a longer intake of breath, a great deal of controlled activity, and intense cooperation amongst the children.

But now we want to go back to the 2s and 3s tables. We draw a long line on the floor and divide it up with small crosslines corresponding to a small footstep apart. Perhaps the line is unnecessary, but we do it just to be on the safe side, for now we are going to try something really difficult! Something we will not actually need until later on, but since we in our class are so clever, we can have a go at it now.

Two children stand at the beginning of the line, one on each side. One walks the 2s table aloud, while the other walks the 3s table, without holding hands at first. Soon they have got it down and it looks really nice, so they try it holding hands. This isn't so easy! But luckily they are good friends and do their best not to embarrass each other more than is necessary. And their friendship holds, if only with the help of the numbers 6 - 12 - 18 - 24, etc., on which they can hop together, while on the numbers

$$2 - 3 - 4$$
$$8 - 9 - 10$$
$$14 - 15 - 16, \text{etc.}$$

they very nearly wrench each other's arms off! With a lot of concentration, however, they keep the rhythm going. The 2s table is the most difficult—in fact, twice as difficult!

In the following illustration a hop is signified by two small lines, and a single line means an ordinary step. One can see the rhythmic alternation of harmony and discord between the two children. If we listen carefully, when it runs smoothly, we can hear a new melody, made up of the two individual tunes. Two contradictory rhythms have created a higher unity.

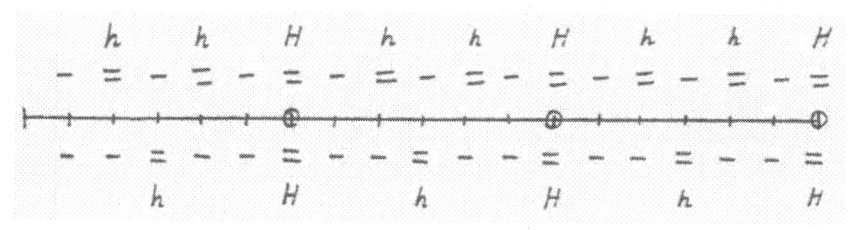

I can jump, I can count, see me jump, see me count.

After this diversion we return to our circle, hold hands, and make a nice ring. As we stand, we do the exercise with which we began, that of swinging our arms backwards and forwards while we count. This time we will see the 2s table emerge, particularly if we are silent when we have our hands behind us. This would look very pleasing to the eye if we could look down from above, especially if the children are not standing too closely together.

Now we ask, could we swing one arm forward and the other back, just as we do when we walk? Let us try and see! First of all, we let go of each other's hands and try swinging them as if we were walking. We all swing the same way—all the right arms forward, the left arms back, and then the opposite—and soon we are walking in place, but only with our arms. It is obvious we cannot *hold* hands, only *bang* hands—as our arms swing past each other in a rush. We can all see this quite clearly. My neighbor must therefore do the opposite of me if we are to have harmony.

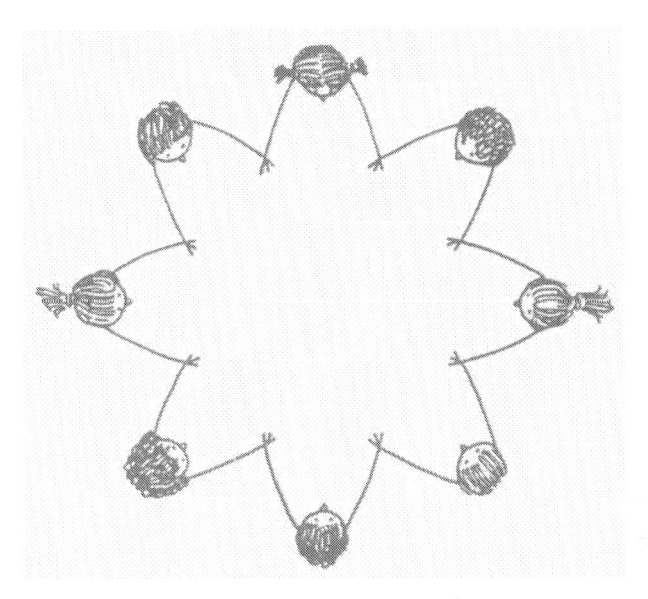

The exercise as seen from above

Already at this point there will be some pupils who are thinking about even and odd numbers. We take for the moment a practical rather than a theoretical approach and choose five pupils, for example, whom we ask to stand in a circle. One begins by swinging his right arm forward and his left arm back, his neighbors join in by swinging oppositely, but when *their* neighbors try to join in, it doesn't work. The exercise cannot be done with five pupils.

When we try it with six, it works, but with seven we have the same problem, and so we hurry on to eight. All of sudden, we find that we can use the numbers from the 2s table. These are the numbers which we call "even" numbers and everybody thinks this is a good name for them. The others are called "odd" numbers, and this too is a fine name!

The best way to try out these even numbers, however, is to swing one arm forward and the other back while counting. It works, whether we hold each other's hands or not. Try it for yourself, and you will agree with the children that it not only looks nice but is fun to do.

Try also imagining this exercise as seen from above, and seeing the children in slow-motion. As adults we can do this, but the children feel it more in their limbs than in their minds.

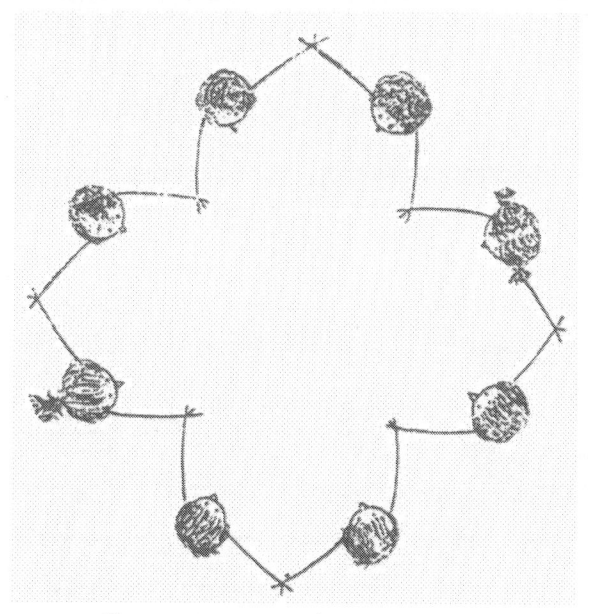

The arms move as if we were walking.

Later on let the class do this exercise again and ask them if it can be done with an uneven number. They will quickly find out that it is possible, with slight modification. One pupil can move both his arms in the same direction, and thus be a missing link between his neighbors. We can put in another "missing link" in another position, and the game takes a new form. In fact, we are once more an even number, and the links are not really necessary.

For the next exercise we will stand in the circle again. One of the pupils steps forward with the left foot first and then with the right, saying at the same time, "1 - 2." The next one says "3 - 4" as he steps forward in the same way. This continues around the circle until we know how many legs have come to school today. The next time round, we put the emphasis on each second beat, stamping, so that the 2 times table is marked out. The same method can be used for the 4 times table, only this time every second pupil stamps as he steps forward on his right foot. This is more difficult, and some of the pupils will step forward with a stamp although the number is not in the 4 times table.

But the real difficulties occur when we try to do the 3 times table in the same way. It will be some time before the class discovers that the problems are caused because a two-legged creature is trying to carry out a tripartite rhythm. Some will not stamp at all, while others will stamp with their right feet and others with their left. The neighbor's foot will instinctively try to help, rather like the way a driver's foot moves forward to brake when he is sitting in the passenger's seat.

This exercise provides the class with a lot of fun. Try it with only a few people, too. The 3 times table with only three people is boring, whereas with two or four pupils, it is more fun.

All these exercises can be modified and carried out with the children sitting at their desks. The 3s table, for example, can be done by clapping hands together on the strong beat and clapping on the desk for the weak beats. Right up through the 6 times table the rhythm remains strong, without really knowing the tables beforehand. Beyond this we need other methods.

Essentially it is the sound which helps to imprint the times tables rather than the children merely learning a list of numbers by heart. It is of great importance that they learn things by heart, but the manner in which this is done is of equal importance.

And it is also important that this way of learning the tables be reflected in the children's books, that is, through visual methods. We will later return to this theme (in Chapter 7), as well as discuss the best way to approach the higher tables.

Until now we have not mentioned where in the curriculum these exercises could be used or even in which order to use them. Every teacher knows from his own experience when the "right" moment comes to use a certain exercise. A great deal rests on the order in which a teacher chooses to teach to his class and also on which books he uses. The author taught for many years in a Rudolf Steiner/Waldorf school and, following the school's philosophy as well as his own conviction, never used a printed arithmetic book, but rather based his methods on the needs of the class at the different ages. He felt it was important that the month-to-month content of the lesson not be determined by the author of a book written some years ago (perhaps even a good book), an author who had no knowledge of his own class with its individual day-to-day needs.

Neither would it be proper here to lay down a specific timing or order for using the individual exercises. It must be the teacher, with his finger on the pulse of the class, who decides when and where. Considering this, it is also obvious that particular exercises should be freely altered to suit the needs of the individual class and situation.

Please, therefore, regard the preceding exercises and also the following as mere suggestion and recommendation which you may put to use in one form or another to activate a class. Remember, too, that they were created, first and foremost, for the children. These youngsters come to school with hopes of getting a teacher, and not, as so often happens, of getting a textbook. This applies especially to the younger children and most particularly in arithmetic.

Arithmetic is noteworthy in being a subject where it is very easy for us to stumble and create fuzzy theories which satisfy the adult intellect. And yet arithmetic is the subject which by its very nature encourages the physical activity so necessary for and so in agreement with the development of the child's will.

Chapter 6
MORE ON TABLES
THE RELATIONS BETWEEN NUMBERS

Earlier we had an example of the relationship between the numbers 2 and 3. We will now go deeper into this theme, which is of great importance in all arithmetic, and explore it in relation to the teaching needs of the younger classes.

When we have reached the point where the class knows the first few tables, we can go into the gymnasium again. 12 pupils stand in a circle, putting their right hands outside the circle. A number-runner walks round the circle clapping the outstretched hands in time to the 3s table. Every time he says a multiple of 3, he pauses while that child squats down. The rest of the class is a choir, chanting and clapping the 3 times table. Normally we would use a 3:4 rhythm for the 3s table, but for this exercise it is best to use a 4:4 rhythm, which gives two beats for the child to sit or squat down.

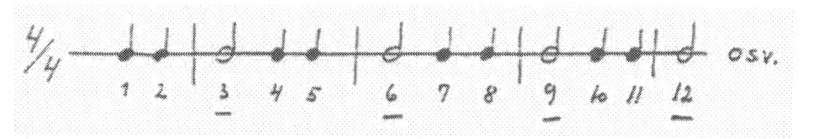

The 3s times table with a 4/4 rhythm.

The number-runner continues round and quite soon four children are squatting down. The class will agree that it looks very nice, and they recognize the figure which has been created.

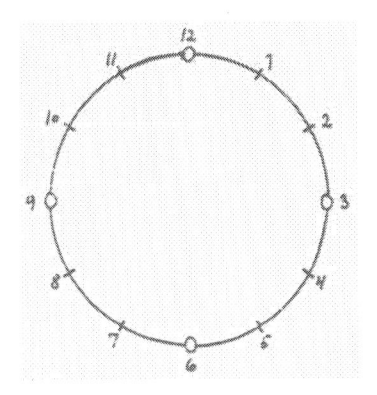

Only 4 pupils will be sitting down.

As the number-runner continues, he returns to those pupils who are squatting, and this time they rise as he passes—so that it is the same pupils who are continually in motion. This becomes boring for the others, so we must try another table.

The 4 times table is chanted to a 5/4 rhythm by the choir, and soon three pupils are squatting in the circle. Again a familiar shape is recognized from earlier exercises.

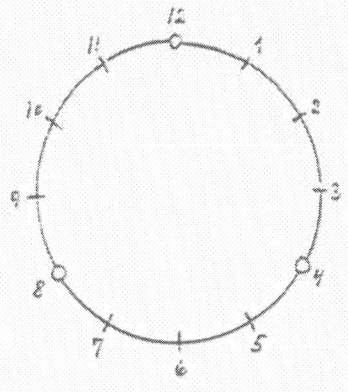

When the number-runner passes a squatting pupil, that pupil stands again. This time new pupils are involved, but it too soon becomes old hat.

Now it is time for the 5s table, and this time we will see something surprising. Try it for yourself, and you will understand the excitement. This time it is not the same few pupils who are involved in all the sitting and standing. To begin with, numbers 5 and 10 sit down, followed by:

$$3 - 8 - 1 - 6 - 11 - 4 - 9$$

and yet we have not returned to any of those already squatting! This seems strange, but it would be even more remarkable if the number-runner actually did come to those three who are still standing, when he has so many others to choose from.

But all will soon have a chance. As the number-runner continues on his way, the excitement mounts. He comes to 2 and then to 7, and finally to lonely number 12, who time after time has seen the number-runner pass right by his nose.

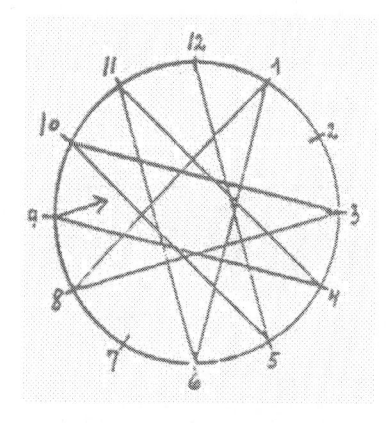

This time everybody takes part in the game.

Everybody stands up again as the game continues in the same order as before, and this time number 12 can relax.

The next day we try the 3 and 4 tables again, but this time we take one of the children out of the circle. The result will be very different from that of yesterday. These two tables will not work with only 11 in the circle. Nor will the 5s table work. In fact, none of the numbers works with the 11 children, so the class decides that this is a very special number. We then try putting in two more pupils, so that there are 13 standing in a circle. Once again, we come to the same conclusion. Numbers 11 and 13 belong to a special group of numbers, and here we promise the children that one day we will explain more about these special numbers, which we call prime numbers.

These games can also be played in another form. Have 12 pupils sit in a circle and give them a ball, which they roll across the floor to each other. Once again, the triangle and square appear. And all 12 of them will be active as before, if they roll the ball to each fifth child. Once they have practiced enough, more balls can be added until the star-shape is clearly marked.

Using balls in arithmetic is a whole chapter in itself. A ball can be one of two things in a lesson, either a nightmare or a source of great help to the creative teacher. A ball can be rolled or bounced. At the same time it makes a sound as it hits the ground and walls and helps our ears hear the rhythms of the tables.

In some of the games, Chinese paper balls can be used. With lots of them, or a very quiet class, we can fill the room with delightful sounds.

The numbers 3 and 4 are closely related to the number 12, and this we can demonstrate to the class in yet another way, which will also bring us to important new knowledge along the way. For this, another game.

On the floor we draw a circle, using a piece of chalk and some string, and then divide it into three as we have done so many times before. At one of the marks, we place two pupils, one on the inside and one on the outside of the circle. The teacher says "one," and the pupil standing inside the circle runs round to the next mark. On the count of "two," he runs to the next mark, and on "three" he is back at the first mark. Here the stationary child has his hand out, which the runner claps as he passes by. Thus we learn the 3 times table in a new way.

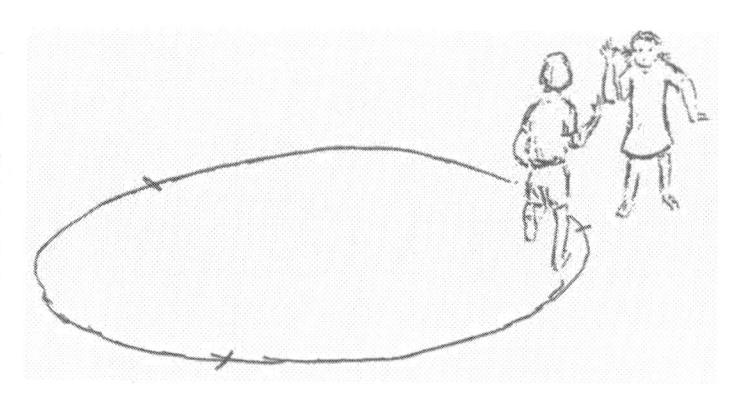

The 3s times table practiced wtih feet and hands

The next child claps hands on the count of 6, as the number-runner passes for the second time. We continue until number 10 has clapped "30" and returns to his place at the back of the queue.

Afterwards we ask the class if they can remember their number. The teacher asks Peter, who was fifth in line, which number he had, and he answers "15."

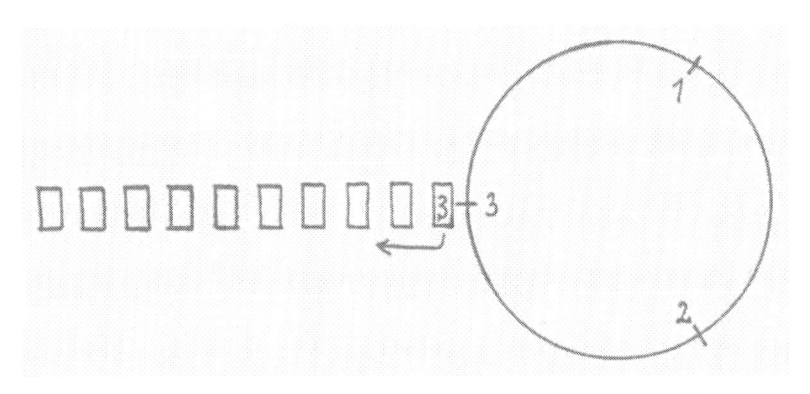

First round problem — second round multiplication problem

"And you, Karen, what was your number?" Karen had number 24 and was number 8 in the line. In this way the class discovers that 15 = 5 x 3 and that 24 = 8 x 3.

Of course the class does not yet know the table "by heart," but it does know how the tables are built up. Before we can begin to learn the tables "by heart," there are many more games to be tried.

We make a serious mistake as teacher if we think that an exercise practiced once is an exercise learned by the class. Hoping that some of the pupils have understood the exercise misses a very important point, namely that repetition, as an on-going activity, is as important to the child as breathing. A child needs both experience and activity, which is why repetition is so important. Thought is always satisfied with a one-off exercise, but a child lives in his feelings and actions where other rules apply. Until we understand this, we will be unable to find workable methods.

When considering the teaching of arithmetic, we must think in terms other than the "once learned, enough" idea, especially if we take the child's unspoken wishes into account. These wishes should make us think differently about both repetition and the child's sense organs. When a young child is very active, he is not merely exercising the muscles in order to develop them. He is also experiencing the different movements and relationship to balance with a far greater intensity than we as adults find possible, but which we can be aware of in the child. Children really "taste" their own movements, experiencing them to the full, repeating them over and over again until they irritate the adults around them to exhaustion. We have hidden memories of such childhood experiences, and the degree to which we can bring them to the surface will determine our ability to teach, especially in the use of repetition as an aid to the development of the senses.

At this point some of Rudolf Steiner's ideas about the senses must be mentioned. He described not just the five senses mentioned in historical times, or the six of more modern times, but altogether a total of twelve senses. In the case of arithmetic, it is not the higher senses related to human consciousness which help us to understand the subject, but instead it is the lower senses which are important.

When we look at modern arithmetic books, they demand all of our intellect in order to understand them. We often hear the phrases: "Mathematics has never said anything to me," or "I could never add two and two." These attitudes stem from so deep within us that in order to understand them we must return to our earliest years when we first learned to stand up, move, and keep our balance, which we all experienced and indeed, continue to do so. Rudolf Steiner spoke about a "sense of movement," for example, through which we experience our bodily movement in all its detail. This sense and the other "lower senses" play a very important part during our early years, and they have a great influence upon our ability to understand arithmetic.

Therefore we must go more deeply and thoroughly into the introduction of arithmetic than is usual. We must appeal strongly to the movements of the limbs and use repetition as our educating principle, rather than as a necessary evil.

Let us reconsider the last game. We walked around a circle, pleased to see the following numbers appear:

3 - 6 - 9 - 12 - 15, etc.

Doing the game with the circle divided into quarters, we get:

4 - 8 - 12 - 16 - 20, etc.

Now let us draw two circles, as shown in the diagram below, and let two pupils stand at the common starting point where the circles meet, but facing in opposite directions. They move off simultaneously in a clockwise direction, in time with the teacher's counting.

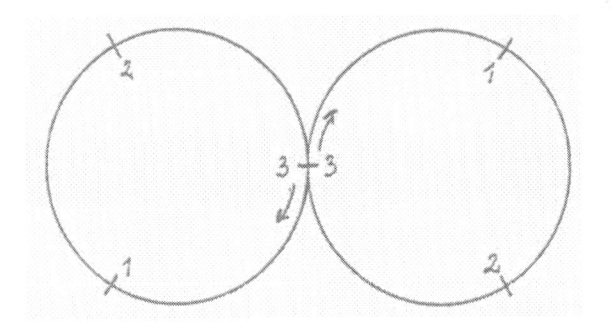

On the number 3 they meet and clap hands at exactly the point where they ought to be. This continues up to 30.

We also try this with the 4s table, and it works just as well.

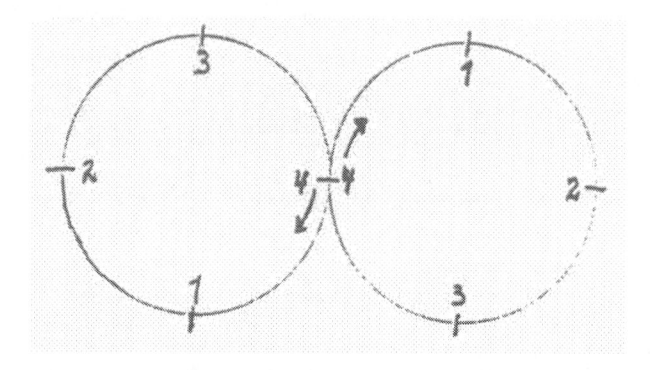

Now let us put the 3 and 4 times tables together, as below.

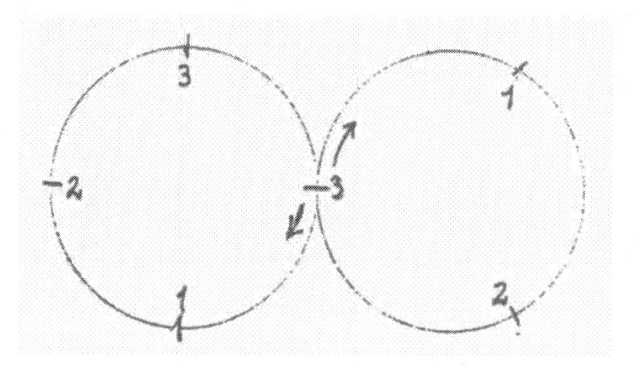

On the count of 3, one of the runners has returned, while the second runner has further to go. Obviously they do not meet to clap hands. On the count of 4, the second runner is ready to clap, but the first runner is already on his way round the circle the second time. On the count of 5, nobody is there. On 6, the first child is there again, but the second child is on the opposite side of the circle. On the count of 7, nobody is there, but when we come to 8 and 9, it looks like they will meet. However, at this point the fours number-runner comes in first, followed by 3. As we reach

10, it seems to the onlookers to be hopeless, whereas on the count of 11 there is tension in the air, and when at last 12 is reached, there is a cheer. If we continue, most of the children will know that the process repeats, and that the number-runners will meet again—some will say right out—on 24.

This means that 3 and 4 are related to and part of a family with the numbers

$$12 - 24 - 36, \text{etc.}$$

As we know from our arithmetic with nuts, there are numerous ways of dividing a circle. We are now ready to introduce many exercises using this knowledge, and so prepare the way for the years ahead. Let us perhaps try the same game with the 3s and 6s times tables. The game goes along without any problems, as do the 2s and 6s times tables.

Now we can try the 2s and 3s times tables together, and we discover that the 3s table must go round twice, and the 2s table around three times, before they meet and clap. This is easy to remember. The 3s and 5s times tables have to walk round a long time before they meet, but the same rule holds true: The 5s table goes round three times and the 3s table around five times.

We now want to know whether this rule always works in the same way. We try the 2s and 4s times tables and discover that the twos runner needs to go round twice before clapping, while the fours runner claps every time he returns to its starting point.

These things are obvious to us as adults, that is, obvious when we think about them, but they are not so for the children. However, these things are obvious to the children's observation, and therein lies the value of the exercise. That which is observed now becomes later a basis for a thinking which is alive and active.

And by "later" we mean, for example, when we reach the fourth grade and are about to begin fractions. It is very useful at the beginning of this difficult subject to be able to say, "Can you remember two years ago when you were only in second grade, and we played a game where we stood in a circle and ..."

They will remember because they were physically active over and over again, some running in the circle, others busy clapping. They will also remember the excitement and the relief of wondering when the runners would meet, and so they will say, "Ah, yes, so that is what you meant ..." The best type of teaching is one which gives the class the chance to say, "Ah-yes" and "Aha."

In the classroom we can play the same game in another form. Two children stand opposite each other. Each child claps his own hands three times and then his partner's hands on the fourth beat, and so on:

1 - 2 - 3 - **4** - 5 - 6 - 7 - **8** - 9 - 10 - 11 - **12**

The same method can then be used for the 3s times table:

1 - 2 - **3** - 4 - 5 - **6** - 7 - 8 - **9** - 10 - 11 - **12**

Now let us play the game with one of the partners using the 3s times table, while the other uses the 4s times table. Sometimes there are no partner's hands to clap when we want them. So we clap in the air, which is quite funny (that is, if the partners stand far enough away from each other). And sometimes the clapping does come together:

12 - 24 - 36, etc.

For this game everybody needs to be very sure of the rhythm, as it is easy to be distracted by one's partner.

When at last everything is running smoothly, we can introduce a more difficult variation. This time four children stand in two pairs, opposite each other. One pair will be told to clap the 3s times table, while the other has the 4s times table. On the count of three there will be a clap, and again on the count of four. It is surprising how quickly they grasp this exercise. There is just enough time for them to take their hands away before it is time to clap again. However, when we reach 12 there is chaos. So, we decide that on 12 - 24 - 36, etc., they will clap out to the sides, creating a circle for the moment. This demands tremendous concentration, but it brings us an exciting new rhythm.

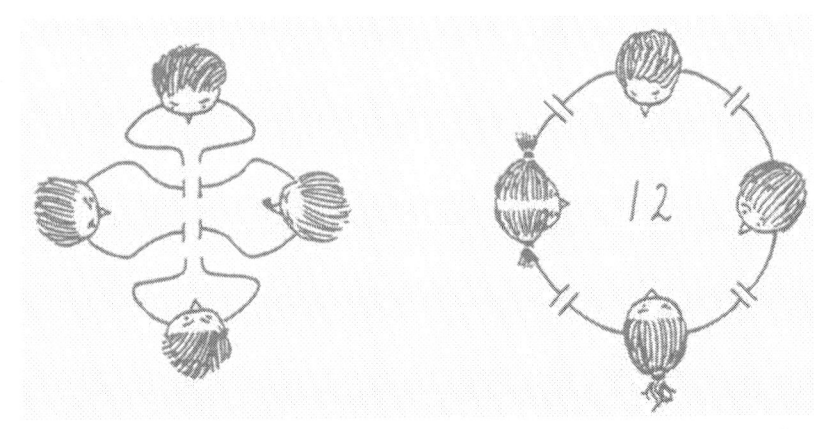

Clapping the 3s and 4s times tables. Much concentration is needed.

We must, of course, move on to the other multiplication tables using this method, but the 3s and 4s tables are particularly good for practicing at first.

We also want to have the whole class working at once, and to do this, we ask them to stand in two long rows facing each other. In this way pairs of children can do different tables at the same time.

While in this formation, try alternating two- and three beat rhythms with each other, while counting:

1- 2 -3 - 4 - 5 - 6 - 7 - 8 - 9 - 10 - 11- 12 - 13 - 14 - 15, etc.

Here we get the 5s times table as well as another shifted 5s times table, starting on 2. If we ask the class to listen carefully to the numbers which we are emphasizing, they will quickly pick out

2 - 5 - 7 - 10 - 12 - 15, etc.

and thereby learn a rule which they will be able to use later on.

When the pupils reach the fifth and sixth years in school, it will not be enough for them merely to see that the tables have a lot in common with each other, but they should be able to work with halves, thirds, quarters, and fifths as well. It is now that we are laying the groundwork for these skills.

We now draw part of our number-runner line on the floor, pointing out, as we draw, that in order to have three spaces we must have four marks. We then position two pupils at each end of the line, as illustrated below.

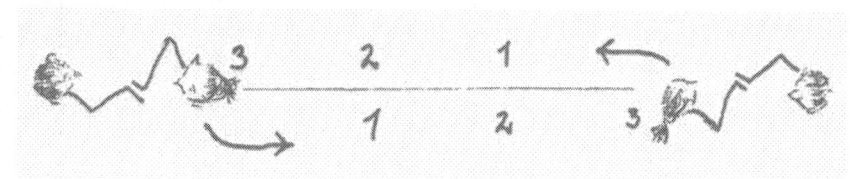

Two pupils learn the 3s table at the same time.

The two "inside" pupils now march according to the 3s table round in the same direction, for example in the direction of right hand traffic (as illustrated), and clap hands with the stationary end pupil when they arrive there. The game goes on and we get 3-6-9 etc. The teacher counts, and the children stop when they get to the end, until the next number is given by the teacher, sending them on their way again. When this exercise has been practiced sufficiently, we change the method slightly, so that the pupils keep moving at a regular pace from one station to the next without stopping. They will therefore reach the number as it is said, which makes the exercise smoother and more rhythmic, as well as more difficult.

We repeat the game using the 2s, 4s, and 5s tables. The pupils are encouraged to be aware of where along the number-line each meeting of the traffic takes place, so that their comprehension of even and odd numbers will be strengthened. The teacher can ask them at which numbers they meet. Those who walk the even number tables answer easily, while the odd table pupils—with the rest of the class—are wondering about it a great deal. But wonder is an excellent preparation for knowledge, so this is all to the good.

It is now time to extend the game as illustrated at the top of the next page. There are three stationary-end pupils, stationed at the outer ends of the lines, but none in the middle. Each line has two runners, one at each end. The game begins very carefully, with the teacher counting and the pupils stopping on each number until the next number is said, slowly as before. Eventually we will have the children walking continuously without stopping. We must remember to keep the right (or left) hand traffic rule. The three pupils who meet occasionally in the middle, place their right hands together as they pass. When they reach the outer ends of their lines, they clap hands with the stationary, waiting child.

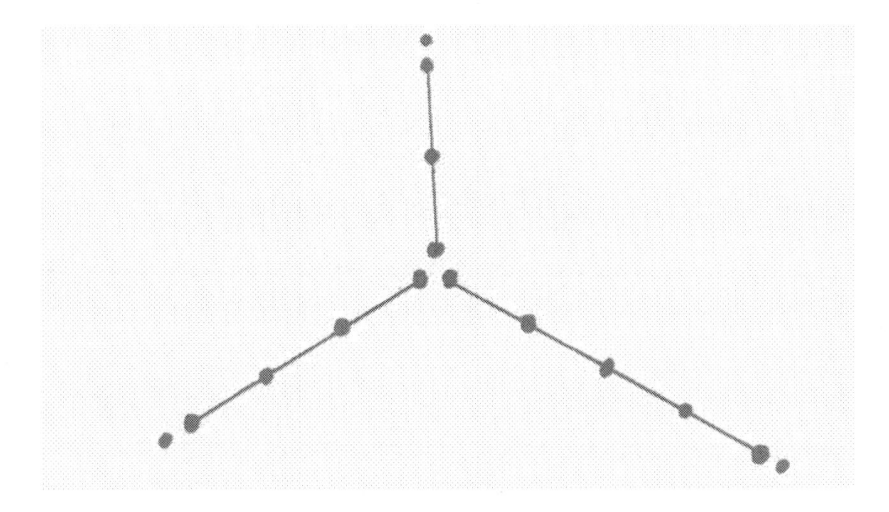

The 2s, 3s and 4s times tables are practiced together, as we prepare for common denominators.

Now we must consider what is happening in the center. On the count of 1, no one is there. On the count of 2, the runner from the 2s times table will be there. On 3, the 3s times table is there, and on 4, both the 2 and 4s times table children meet. On 5, no one is there, but on 6, 2 and 3 will meet. On 7, again no one is there. At this point, we are all wondering if the three runners will ever meet. We continue to play and discover that at 12 they do all meet, once more.

It is now time to bring in the 5s table along with the 2s, 3s and 4s tables. The children try to work out what this will mean to the game, but they don't get very far just thinking about it, so we must start the game and see what happens.

This is going to be a long and complicated process, but it will not be boring, because there will be plenty of drama. When 24 is reached, there is great excitement, and everybody turns to look at the 5s times table, who

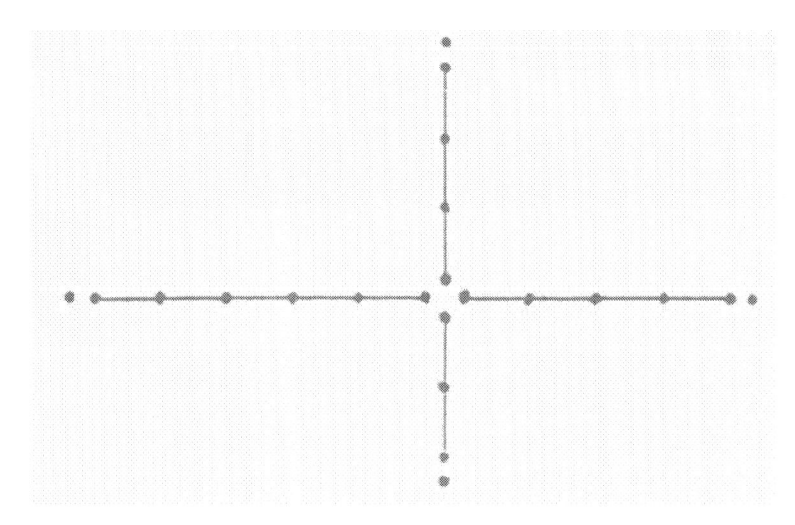

The 5s table has been added, and it does not make things easier.

is spoiling everything because he is one station behind the others. When he reaches 25, he is home, but all the others have gone. When the number 30 is reached, it looks a little better, but the 4s times table is away from home. When the runners reach 36, once again the 5s child lags behind. On the number 40, the 3s child is left behind, which is unusual. At last we reach 60, and all our worries are over.

This game can be simplified by using the form (top of next page), which also is of advantage for other purposes and subjects (for example, the planetary orbits).

The children are familiar with the circle for practicing the tables. We place the children concentrically, each with his own circle, and the starting points all in a straight line. On the count of 1, the straight line is broken. The class now wonders when the straight line will return. The problem is exactly the same as before, with a few variations, and gives us further opportunity to learn about numbers.

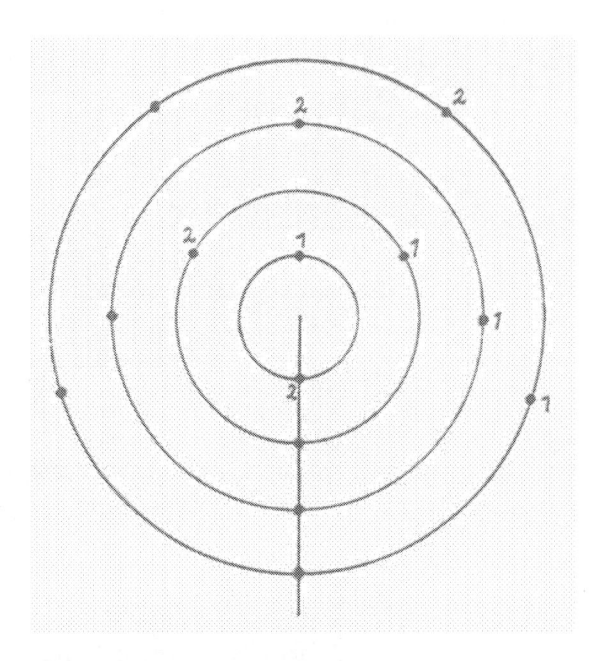

Four tables in their daily orbits

The prime numbers after 5 awaken the children's interest because they leave the starting line empty. We keep a close watch on the starting line and discover, apart from all theory, that we must watch these irregular numbers carefully. The starting line is empty when a prime number is called, and, if looked at in the right light, so are the numbers themselves.

Let us consider for a moment other aspects which should always be in our minds when working with children. It is not unimportant whom we place where in the circle. If a lively, sanguine child is placed in the center, he will enjoy turning round and round, and it is doubtful that he will tire of it. If he does tire before 60 is reached, it will not hurt him to continue. This position, however, would not suit a phlegmatic, who would benefit from being placed in the outer circle. That position would also suit our thoughtful melancholic, who, while revolving around the other circles, can oversee the whole game. The choleric needs to be positioned near the sanguine.

Try asking, before the result 60 is known, if we will ever manage to reach home all at the same time. Our sanguine will not bother to answer because he already will have moved on ahead of all the others, if we are not in strict control. It is likely that the choleric has also moved on, trying to take two steps at once to get the whole thing finished, or perhaps giving those in the different tracks a push.

The phlegmatic will continue to go round and round, counting and counting, and preaching for calm and order, while nobody listens to him. Our melancholic may eventually decide that the only answer is to go backwards to reach zero.

During all this activity, the children will experience many things of use later on, a knowledge: of prime numbers, of common denominators, and of smallest common denominator. All these experiences are seeds for the "aha" experience.

This circular form of the game is very useful for later on when the children are learning astronomy and to understand the conjunction between Saturn's long, slow orbit and Mercury's sanguine revolution about the sun.

We can also take notice of how the pupils stand in a straight line at the beginning of the game, but as soon as movement begins around the circles, the line curves and eventually becomes a spiral. This can be seen more clearly if the children hold a long piece of yarn or string between them, unwinding it from a ball as they go. The effect will be even more clear if we also use higher tables, for example the 5s, 6s, 7s and 8s tables, instead of the 2s, 3s, 4s and 5s tables.

Through experiences such as these in the mathematics lessons, we can develop themes which satisfy a deep-seated need in every pupil, namely that the different classes and subjects in school are related and connected with each other, and all show different sides of a whole.

Chapter 7
DRAWING AND ARITHMETIC

In this chapter we will look at drawing exercises that can be used to help in the understanding of the multiplication tables.

We ask the class to stand in a long line. They count out loud down the line, starting with 1, and those who have numbers from the 3s times table take a step forward. The whole class looks and studies the pattern which can now be seen.

o o   c c   o o   o o   c o   o o

  o   o   o   c   c

The same exercise is repeated using the 4s and 5s times tables.

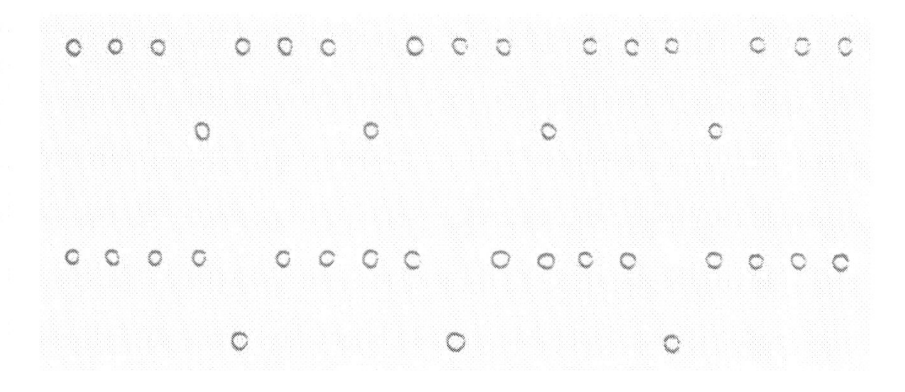

o o o   o o c   o c o   c c o   c o c

   o   o   o   c

o o o c   c c o c   o o c o   c o o c

    o   o   c

As we move through the tables, the distances between the pupils that have stepped forward grow bigger and bigger, and the groups table, these groups consists of 4 pupils, in the 4s times table 3 remain, and with the 3s times table only 2 pupils remain behind.

In the next exercise the whole class, except for a number-runner, stands in a long row. The number-runner then runs in a half loop from the first number of a chosen multiplication table to the next and so on:

He skips over every other child.

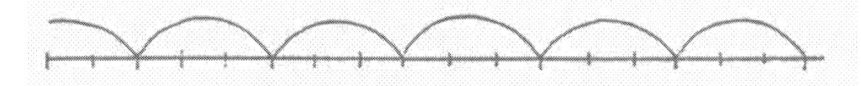

Now he skips over two children every time.

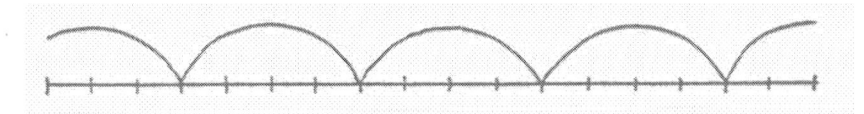

And now three each time.

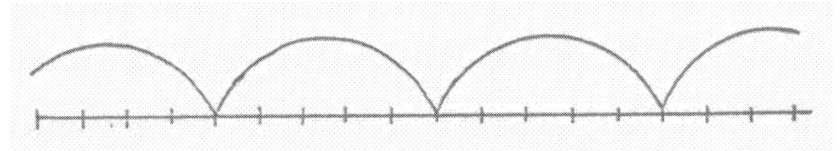

Finally, he skips four each time.

It is important that the class experience this as intensely as possible. We continue, therefore, with another exercise and draw a line on the floor, marking off the numbers evenly and clearly marking the 5s and 10s times tables.

One pupil starts at the 0 point and is instructed to walk with a 3-unit-long step down the line. He knows the 3s times table and uses the markings for 5 - 10 - 15 - 20, etc., as reference points. But we want to get the 3-unit step into the blood, which is why we practice it over and over, as well as other tables, even though the children do not yet know them by heart. The pupil is instructed, for example, to take five 3-unit steps forward and see where he ends up. He is actually learning that

$$15 = 5 \times 3.$$

Many such examples from different tables can be practiced. Later on, we can blindfold the children and see if they are able to land on the right number. For example, if the teacher says, "I want you to take four 5-mark strides, but first tell me, how far will you go down the line?" The child answers, and then he is blindfolded. He strides down the line, and when he stops, the teacher removes the blindfold so that he can see how near he is to his target. It was probably not too good; perhaps he has come too far down the line, so he must try again. This time, he stops short of the mark, but at last, after a few more attempts, he manages to get it more or less right. It is rather like the story of the blind Polyphemus, who threw stones at Odysseus. The first was too far and the second too short, but Polyphemus was allowed only two attempts.

Meanwhile, the rest of the class is watching with great interest and they are learning far more this way than we can possibly imagine. Later on, everybody will have a chance, and our abilities in arithmetic will be stretched to their limits.

What you do not yet have in your head, you can have in your feet.

We also try some of the larger numbers, telling the class, "These tables really come later, but let's try them out now just for fun." If we take, for example, the 11s table, the class will quickly discover that the number-runner cannot jump so far. But this can be overcome if he is held up on either side by a couple of his friends, so that he swings through the air.

11 - 22 - 33 - 44, etc.

Another way to overcome the problem is to mark off the distance by running in a long arc, from one number to the next, e.g., 11 to 22, to 33, to 44, etc.

With the larger, even-numbered tables, it is possible to have a hop midway, then land on the table number with both feet together—at least we can try it. It is not as easy as it sounds. We will then learn that the 12s times table (e.g., 12 - 24 - 36) is actually every other number in the 6s times table. There will be those in the class who will try the same method with the odd-numbered tables, and this will give them something new to think about.

One way to take long steps for the larger number tables

There is yet another method which always awakens enthusiasm in the children. Get a set of vaulting poles for the class, and let them use these to practice their tables. Like a pole vaulter they will be able to jump far and also land on the right number. On another day use a hopscotch marker, tossing it ahead to the next number in the table. They will find this a little more difficult, but their powers of observation will be sharpened. With such a marker stone and a line of numbers, it is also easy to practice addition, as well as other aspects of arithmetic.

Count and jump.

They can now illustrate these exercises in different colors in their books, as below:

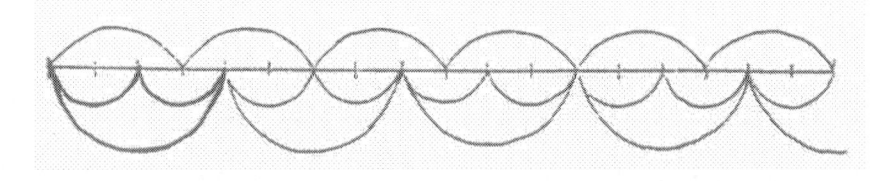

Once again the children will experience—this time visually and manually—how certain numbers are brought forward, while others are left out. If we take, for example, the 2s, 3s, and 5s tables together up through 30, the numbers marked below will clearly stand out as being skipped over.

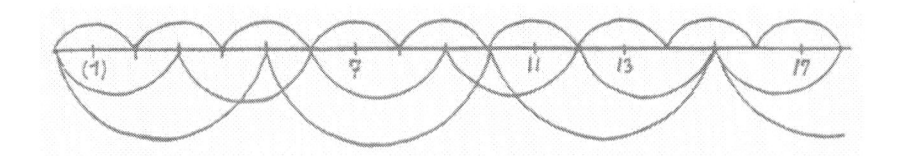

It is important to practice the bigger tables more and more in this way as we go along. These exercises have been designed primarily to help the children experience the progression and distance between the series of numbers within each table. Measurements using 3 and 4 are easy to deal with, but with the bigger tables, larger measurements are needed. We discover that our legs are not long enough, so we use "wings," but our wings soon tire, and we become aware of the need to find a better method.

Next we discover that our eyes can jump very quickly from one number to another, and finally, we find out that the quickest method is jumping with our thoughts. Thus, we have discovered a method which has no limits. We now know that we can jump quickly from one number to another, as fast as we want and as far as we wish. But we also learn that we must jump very carefully; it is easy to make mistakes moving so fast, especially with the bigger numbers!

It is very important that the class actually experience this—how we can reach far very quickly. Ask the ten children who are taking part in the game this round to line up along the wall. Count slowly to 10. The first pupil takes only one step forward because he is the 1s times table. The second stops on the count of 2, for he is the 2s times table, and so on. Only the last pupil walks all the way, one step for each number called up to 10. The result will look something like the leftmost diagonal line in the following chart on the next page:

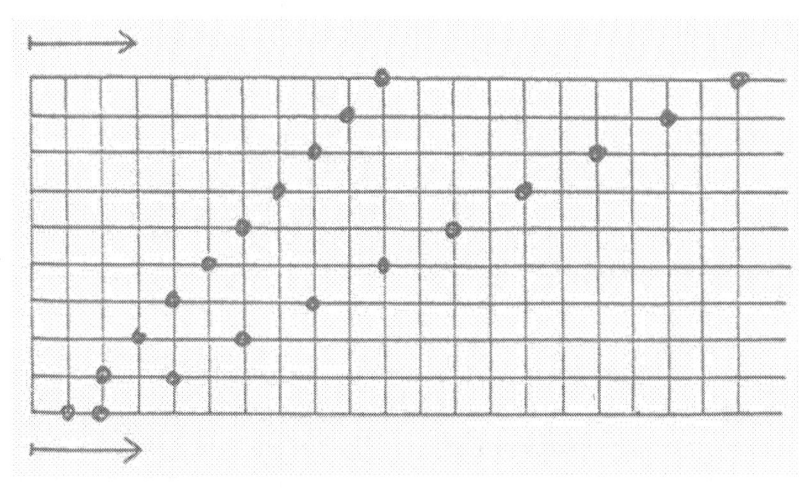

It is useful to have this kind of board on the wall. For the younger classes it should be drawn on the floor.

These places correspond to places they would have had in the earlier exercise using measured strides—the child standing out furthest would have taken a 10-mark stride.

The teacher counts up to 10 again, and the children move forward. Every pupil has now moved forward twice in his table, producing the second slanting line of the illustration. We can now go back to the beginning and do it again but a little differently this time. When the teacher says "one," all the children are to move quickly forward to their first halting place, that is, the first diagonal line, and on the number 2, to their second stop, and thereafter according to the same principle.

The children can see how one side of the room moves very quickly while the other is much slower. The principle of the "measuring stick" is understood by some children at once, but they need more practice, and they should all experience what it feels like to be positioned on the left, right, and center of the line.

If it is a second or third grade class, the teacher should now tell the children the story about the Hare and the Tortoise, and then let them dramatize it. To do this, we take the two main characters and let them practice "hare" or "tortoise" steps across the room. When the children are the tortoise, they must take tiny steps, but when they are the hare, they take very big hops. Let a hare and a tortoise begin from a wall and hop outward into the room, hopping with both feet together and hopping at the same time, even though the hare's hops take him further. After all this hard work, the class will know much about the difference between the 2s and the 10s tables. They will also appreciate why the long-legged hare got so tired.

At this point it would be a good idea to let the children draw pictures from the story, showing the hare and tortoise at different stages in their race.

Again and again we practice—both large and small steps. For the large ones we may need "flying" help. We try again with our eyes blindfolded to reach the right point and see where we end up—and continue until we have more or less got it right. This exercise gives the children a good feeling for spatial relationships. We then practice again without the blindfold—they will soon be well-oriented on the positions of the numbers: 6 lies 1 after 5, 9 lies 1 before 10, 12 lies 2 after 10, etc.

The transition from feeling to orientation, that of clearly knowing where we are going, is of the greatest importance. Feeling lies so much deeper within ourselves, so to speak, than that which we have called orientation, or knowing. Here we find support in Rudolf Steiner's teachings about the senses and say that feeling-perception given us by our movement relates to the lower senses, orientation judgment-knowing to our higher senses. A more detailed study is outside the scope of this book, but reference can be made to Steiner's work.

This development from immediate, feeling-perception to orientation-evaluation has a continuation. With the children's help we build a device, illustrated below, of the type used in olden times to measure fields. We, however, will use it for measuring along our number-line, where we used our feet before. We have now reached a third stage. After "feeling-sensing" and "evaluating-judging," we now have a technical device which we have put outside of ourselves and which is no longer connected to our bodies, but which we can nevertheless make good use of for outer measurements.

The number-line which we draw on the floor must now be exact. This causes no problems, because we have now reached the third grade, and here we are learning about weights and measures.

The multiplication table measurer at work

Here for the first time we experience tables from a "technical" viewpoint. Our new apparatus does not make mistakes; it is completely stiff-legged and cannot have very many sensations, nor need it take time to orient itself within the room.

It is now time to do some comparisons, using a second, identical measuring device. The first "measuring-stick" is set for the 4-mark stride, while the other is adjusted to the 2-mark stride—both of them must be very exact. The first child "walks" his stick 3 steps forward and the second child has to "walk" his own stick forward 6 steps to catch up with his companion. This is something which we all knew would happen, and now many similar exercises can be developed. For example, how many strides must big brother take to meet up with little brother who has taken 10 strides? Or what would happen if little brother had taken only 5 strides? Ah, yes, that would mean that big brother would either have to go too far or stop short. Thus we learn that as little brother moves, big brother can meet him occasionally. Little brother can stop anywhere; big brother sometimes seems more clumsy. We let the children draw all these exercises in their books, not the least so that they can show that they are aware of the direction of the handle, for each turn of the device.

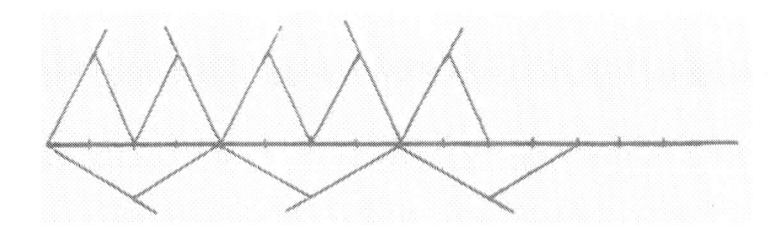

Next we will take 7-mark strides together with 3-mark strides and ask ourselves whether the two tables will ever meet. We find that they meet on number 21 and again on 42.

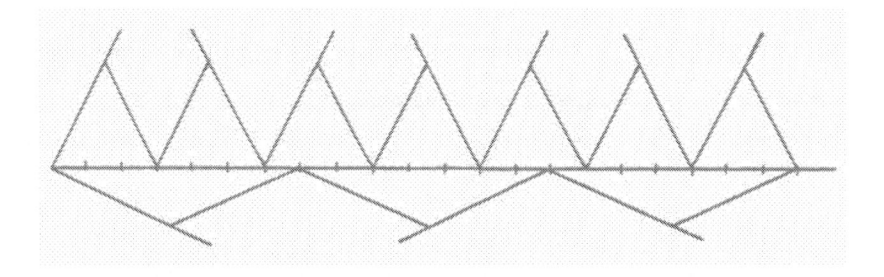

Help the children to see the symmetry of the figure, ignoring the direction of the handles. If we want even better symmetry we can use 7- and 4-mark strides to meet at number 28.

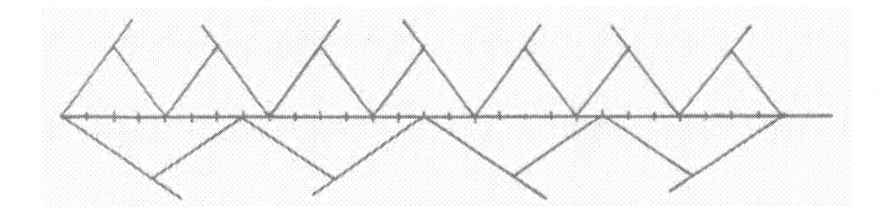

And if we want perfect symmetry, we will have to continue onward—in both cases two full periods. Try it yourself!

This simple device opens the door for the children to a wealth of exercises—and they may be of a very practical nature—which they would surely miss no matter how many "real-life" examples with trucks and tile floors we give them. Besides, our measuring instrument is something which we have made ourselves and always have there in the classroom, something we cannot say for tiles and trucks.

Now, let us place one of the pupils at 0 on the number-line, another at 50, and adjust the instrument for 8. Is it possible to reach the number 50 with the measurer? If not, how close can we come? Having tried it, we draw the result on the blackboard and also in our books.

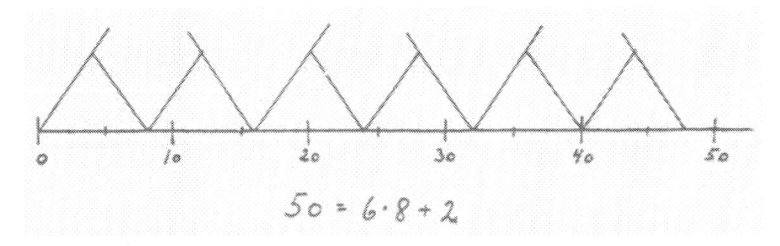

$$50 = 6 \cdot 8 + 2$$

It is obvious that we cannot reach 50 with one instrument, but we ask ourselves, is it possible using both at once, both set for 8-units, moving towards each other from each end? No, we will always end up with a gap of 2 however much we go backwards and forwards from both ends.

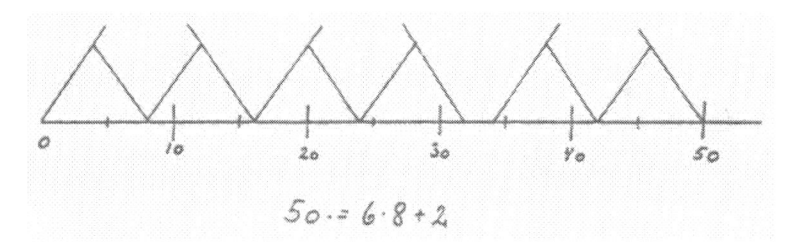

$$50 = 6 \cdot 8 + 2$$

We can also get the situation where we have an overlap of 6.

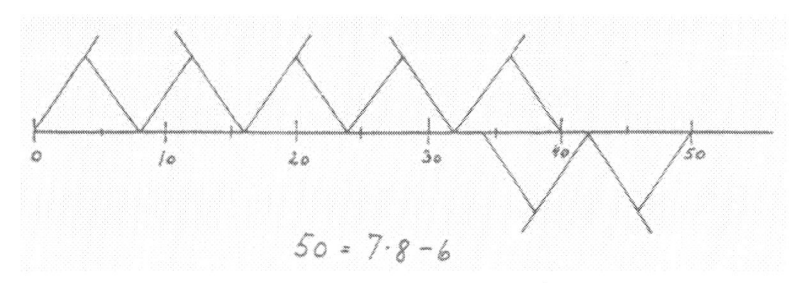

$$50 = 7 \cdot 8 - 6$$

107

If we set one instrument for 8 and the other for 3, will they ever meet? Suddenly the class is in an uproar! Here we must remember that everything which is worth anything begins in chaos, and that our task is to try to form a cosmos, an order, out of it.

The two children experiment, going backwards and forwards several times. They attack the task with feverish enthusiasm, which means that their work lacks precision, which is why they have to keep starting over again. Eventually the teacher has to intervene and help them cooperate more with each other. Some in the class will realize how important it is to know their 8s table, so that they know exactly where they are going. Thus, they begin to discover that thinking about a problem first leads to confidence in action.

But this aspect of problem-solving, too, has to develop at its own pace.

So we continue with our measuring sticks, and we find that upon reaching 48 the 8s times table is very close to 50. So we let the 8 retreat with the 3s times table following close behind from 50. After the 8s times table has made only 2 strides back (to 32), the onlookers cry out, "They are meeting!" Thus, we see that 50 equals 4 strides of the 8s times table plus 6 strides of the 3s times table.

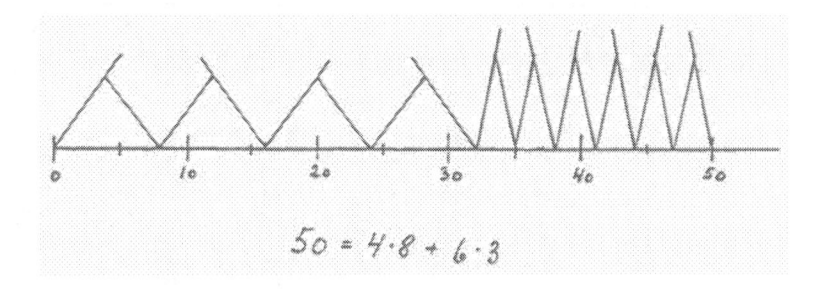

$$50 = 4 \cdot 8 + 6 \cdot 3$$

This is a very difficult concept for the class, but they love working with the instruments.

What if instead we let the 8s times table hold back at the beginning, start at 0, and push the retreating 3s times table towards 50? Again a difficult task to work out and also to draw.

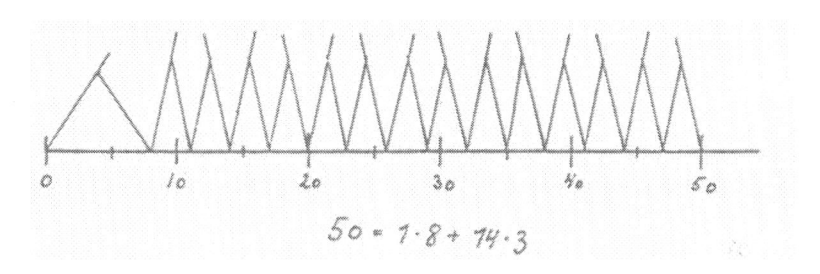

$$50 = 1 \cdot 8 + 14 \cdot 3$$

This is a game which always arouses interest. It appears to be a fight between two opponents, but this fight is really all about their interaction, coming to terms with each other. It can be performed with many variations and with varying degrees of difficulties. The game also contains elements that come up later in mathematics, over which some of the pupils may be wondering now. For example, in the following table, the first and fourth columns give much material for thought.

8	16	24	32	40	48	All the numbers from the 8s table
42	34	26	18	10	2	Are there any numbers here from the 3s table?
50	50	50	50	50	50	

Other aspects of this game involve equations in two unknowns, but are too advanced now for the third and fourth grades.

In order to prepare for the next exercise, tell the children to draw a number-line and mark off the 5s times table. Then let them walk along it using another table, e.g., the 6s times table.

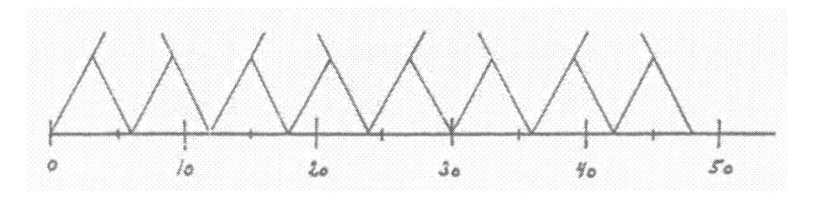

The 5s and 6s times tables meet.

Children can learn a great deal from dividing a line up into 10 even spaces, without using any measuring device. First of all, they must find the middle, then divide each half into the ratio 2::3, and finally add the other marks.

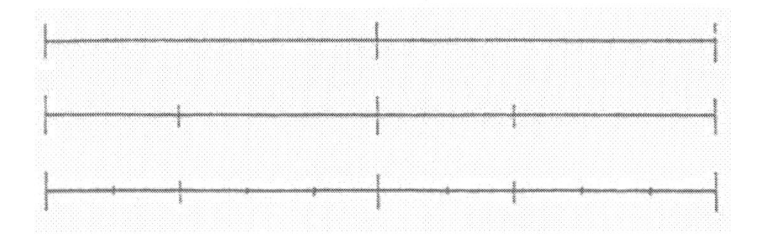

The two previous exercises are preparation for the next problem. First we observe the movement of the 6s times table along the number-line and discover that the first stride forward has already brought us 1 beyond the number 5. The next has brought us 2 beyond the number 10, and the next brings us 3 marks past 15. With the measuring stick we experience immediately that this has to be. It is one of the many obvious facts, of which mathematics abounds and toward which we tend to teach children to "think" their way (in order to train their intellects)—without investigating whether it is utterly certain that this is the best way. It is better to allow children to follow in the footsteps of all the great mathematicians: first experience and intuition, then the use of thought.

Obviously, the 6s table is always going to be ahead of the 5s table. The 5s table is rather like the lines between pavement stones, and the 6s table like our own big strides. Our strides are a little bigger, so that after we have taken five of them, we have accumulated one whole pavement square extra. We have learned a rhythm on our little walk and ought to be able to illustrate it in one way or another. This is where the exercise of dividing a line into 10 equal parts comes in.

The class draws 10 long lines underneath each other and divides them into 10 equal parts, so that we have marks from 1 to 100. We must now find the marks which belong to the 6s times table and put a colored ring around them.

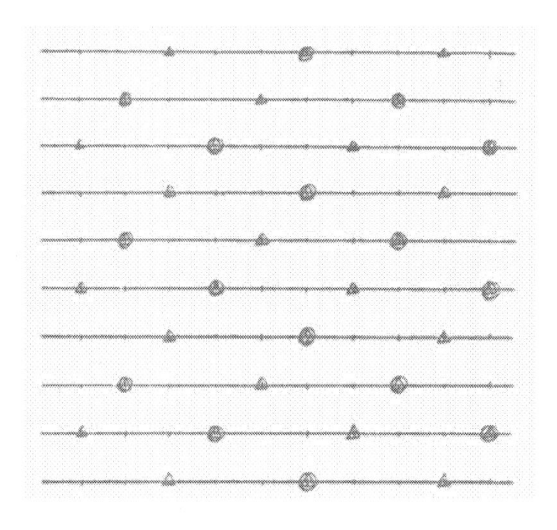

Each table has its own line structure.

We can see how closely related to one another these two tables are and perhaps also how they help each other to bring forth the 9s times table.

Other tables give different structures, and each one shows us something of the nature of that particular table.

It is a good idea to let the children make their own diagrams, but we must be careful not to put too many tables on one drawing, so that the lining up pattern disappears. But if we do put on too many one day, for example, all the tables from 2 to 10, this too can teach us something—the numbers not circled will in fact be the prime numbers from 10 to 100. We can write these numbers down, giving us a list of the prime numbers—another table of numbers (*tabla* in Latin means *slate* or *board*, for writing things on). This table is not at all regular and orderly—altogether different and much more difficult than the ones with which we have been working!

So far we have been working with numbers from two points of view. First, we tried to describe them as numbers of the kind that young children intuitively experience. We called these numbers "essence" numbers, although we struggled for a better name, and gradually gained a new understanding of the qualitative content of numbers. We assumed that numbers of this kind actually exist for children—because we have so many examples of how people of earlier times experienced them in this way and also because we see a parallel between mankind's early history and the development of the child's soul.

We have already seen that these numbers were most nearly related to our cardinal, or quantity, numbers. But they differed by having a qualitative nature, which in a way seems like a paradox to us, because we are used to seeing only the quantitative in numbers. We saw these numbers in relation to organic phenomena where the associated number appears as if of necessity. The number was the way it was because the organism would otherwise cease to exist as an organism. For example, to the young child, his mother and father represent an organic number 2, a pair, which is not arrived at by counting, but by being. The beak of a bird, with its lower and upper jaws, is another organic two, not by virtue of counting but by virtue of being a beak.

Another example is the function of the hands. At some time or other we have all tried to carry a large number of small items from one place to another. For example, we want to carry some cups, plates, milk bottles, knives, forks, plus the leftovers from the meal, or perhaps, a large bundle of wood to be burned in the fireplace. Some of the things move, and we loosen a couple of fingers to try to stop the objects sliding out and making a mess on the floor. Pause in such a situation for a moment and observe what is happening with your left and right hands. See the marvel in what they are doing! Here is a general-purpose function the likes of which cannot be seen anywhere else! This function has to do with the number 5, but as we said, apart from all counting. Afterwards we can count, observe we have come to 5, and therefore determine the number of fingers to be five. The quantity 5 did not exist earlier, in the beginning, only the function of the hand—the one function. And out of this one function arises the organ, the purposeful whole. We can count because our organism has built-in rhythmical capabilities, and through counting we can come upon the quantity, in this case the quantity 5.

Our language gives us these two (or three) words: number and quantity or amount, which we may use to differentiate between the two concepts. The word "number" can be associated with the unity of an organism. But here the number must be understood as related to the structuring power which divides the entity into subordinate functional areas, which we as observer then manifest in our consciousness as a number experience. There are no "quantities" in nature; these exist only as creative acts within ourselves. Nor does nature have any "numbers," except in the sense of subdividing powers as mentioned above. Some parts of the new mathematics, therefore, make a major error when stating that a number is a characteristic of a set in the same way that color is a characteristic of an object.

Numbers never exist apart from a human being. If we must use the expression, then we would necessarily have to say that numbers are characteristics of us human beings. But organisms themselves are whole and have different functions, and one way of describing this phenomenon is to use number concepts. Numbers have to do with activity. This activity expresses itself outside of ourselves as a subdividing into different functions, within ourselves as that which we normally understand by "number." The number we experience as such is not be found outside of ourselves as a characteristic of a set.

If we take the time to study entities outside ourselves with care, we will experience the qualitative numbers. If we then let our fingers begin to roam with the help of our rhythmical abilities and "walk" around, let us say, the rays of a sea-urchin, through this rhythmical movement we come to a certain number. Here we have an ordinary cardinal number, and it is the result of a bodily movement of one kind or another—with the fingers, with the feet, or perhaps with nothing more than a jump of the eyes. These cardinal numbers are entirely our own creations; they do not exist in the surroundings. This is the one place where we can be in agreement with those philosophers who would say that the world exists only in our own imaginations.

We therefore need to differentiate between three phenomena: the pre-cardinal numbers which we experience as qualities, our rhythmical counting numbers based on these, and our cardinal numbers which are no longer bound to the qualitative but have gone over to the purely quantitative.

The first two groups have been discussed in the preceding chapters, and we will now look at the third.

Chapter 8
NUMBERS AS QUANTITIES

For children, all learning must begin with movement. Let them point at their fingers as they count 1 - 2 - 3 - 4 - 5 - 6 - 7 - 8 - 9 - 10 and discover: "I have 10 fingers!"

A basic, archetypal picture of a number is combined with counting, and we thereby reach a quantity, a measure of how many. They can also count their toes and learn that they have ten.

The hands have helped them to comprehend—to grasp mentally—the feet have helped them to understand. Comprehension and understanding are the final links in a development which begins with movement.

Follow this by letting the children take a handful of nuts and count them, placing them one at a time in a pile on their desks. Movement brings them forward to a cardinal number: 12 nuts, for example. Make sure that the children make large, deliberate movements—this is not a waste of time; it is the children's natural way.

When people comment that during the time spent on these movement exercises the children could have done many, many sums, it is because they have not realized how many sums these exercises are preparing the groundwork for—nor how few of those sums the children would actually be able to do at all, had they not been fortunate enough to avoid such an "earlier-the-better" start.

So the children have placed 12 nuts in a bunch on their desks, and we now ask them to try to make a new grouping, this time in a squarish form, more or less.

From chaos we find order. We ask: Are there other ways of doing this? Yes, of course there are!

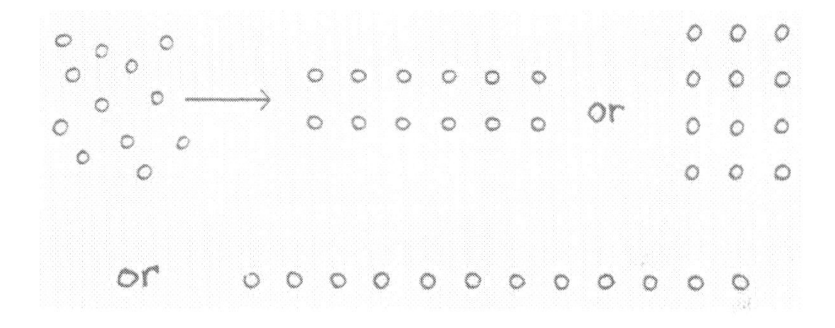

Next, we leave the classroom and go into the hall. All 25 children gather together at one end of the hall, and one child is asked to arrange the rest of the class in a nice, orderly fashion at the other end. One by one he directs them where to stand and soon it looks like this:

117

Each row holds hands and says aloud, one row after the other:

"We are four."

"We are four," etc.

The teacher then says,

"24 is ...?"

and each row continues

"4" + "4" + "4" + "4" + "4" + "4"

and at last everybody says,

"24 is 6 x 4."

The teacher then says "change" (or switch), which the children all know from watching marching bands, and the class now looks like this:

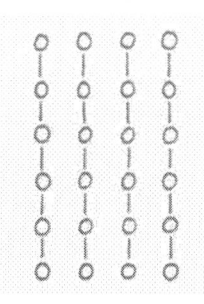

They then repeat one after the other:

"We are six."

"We are six," etc.

"24 is '6' + '6' + '6' + '6.' "

"24 is 4 x 6."

The 25th pupil now has the job of rearranging the class in some other order.

This time the class says:
> "We are twelve."
> "We are twelve."
> "24 is '12' + '12.'"
> "24 is 2 x 12."

Once again the teacher calls out "change," and the class then stands like this:

```
o o o o o o o o o o o o
| | | | | | | | | | | |
o o o o o o o o o o o o
```

to which the text is:
> "We are two."
> "We are two," etc.
> "24 is '2' + '2' + '2,' etc.
> "24 is 12 x 2."

Once again the teacher says "change," the pupils turn to find their new positions, with some of them spinning all the way around (!), and the result is this simple form:

o-o-o-o-o-o-o-o-o-o-o-o-o-o-o-o-o-o-o-o-o-o-o-o

The children answer:
> "We are 24."
> "24 is 24."
> "24 is 1 x 24."

The teacher calls "change" for the last time, and each pupil stands on his own.

ooooooooooooooooooooooooo

This time the call will be:
"I am one."
"I am one."
"24 is '1' + '1' + '1,'" etc.
"24 is 24 x 1."

In this last round there is a change in the tone. We can hear the individual characters. There are the strong voices who want to do it all alone, the shy voices of those just gaining self-confidence, the class clowns who are having so much fun learning to stand on their own two feet, and the sleepy voices who are learning to be on time.

The next day, we go into the hall once again and ask the class how many we need to make a square with the same number of people on each side. Immediately, one of the pupils answers "nine," which we then try.

o—o—o
o—o—o
o—o—o

Remembering the importance of repetition, we repeat yesterday's ceremony:

"We are three."
"We are three."
"We are three."
"9 is '3' + '3' + '3.' "
"9 is 3 x 3."

To the great joy of the class the teacher calls "change." But it is almost the same! Some in the class discover that it is not exactly the same, since they are holding hands with different partners. But if we look only at the form, yes, it is the same. And this movement towards pure numbers is good to have experienced and to be able to return to later.

Now we ask if there is a smaller square that can be made, with an equal number on each side, and find that 4 is the answer.

○ ○

○ ○

We ask the class, "Is there an even smaller one, such that if we say 'change', nothing will change?" Soon someone comes upon the idea of a single person standing by himself.

Tom volunteers. He stands up, takes his position. The teacher says "change," and nothing happens—the picture looks just the same! "Well," we note aloud to the class, "Tom isn't really a square, so what could we have expected, but since 1 is truly 1 x 1, let us pretend for today that Tom has a width of 1, looked at from any of his sides."

Finally, we try making a square with 16 and one with 25. With only 25 in the class we cannot go any further.

On the other hand, we can try to place our 9 pupils in interesting ways:

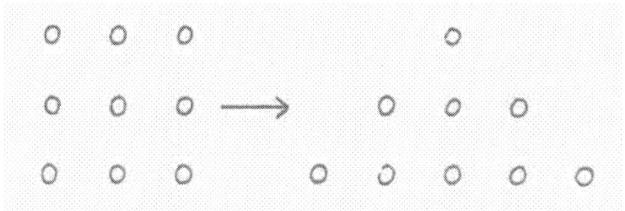

```
o o o              o
o o o  →     o o o
o o o      o o o o o
```

And then with 16 pupils –

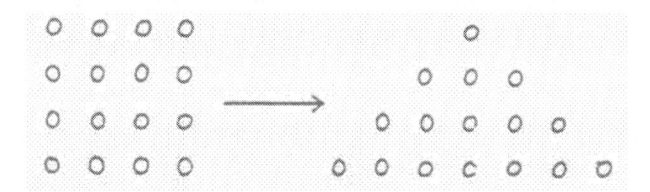

```
o o o o                o
o o o o            o o o
o o o o    →     o o o o o
o o o o        o o o o o o o
```

And then with 25 pupils –

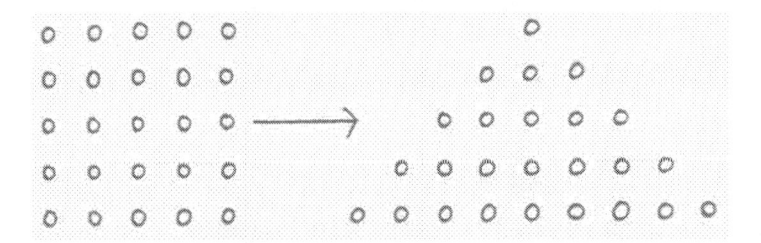

```
o o o o o                    o
o o o o o                o o o
o o o o o    →         o o o o o
o o o o o            o o o o o o o
o o o o o          o o o o o o o o o
```

Oh, yes, we forgot to try with 4 pupils –

```
o o              o
o o  →       o o o
```

122

We then repeat the whole exercise in our books and discover what it is we are really doing. For example, when we have 9 pupils –

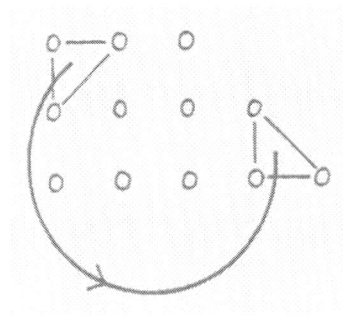

In the same way we can rearrange the other figures. One day, when we have plenty of time, we return to this exercise. Looking at the triangles and counting line by line, starting from the top, we learn that

$$1 = 1$$
$$4 = 1 + 3$$
$$9 = 1 + 3 + 5$$
$$16 = 1 + 3 + 5 + 7$$
$$25 = 1 + 3 + 5 + 7 + 9$$

In the square with 1 pupil, 0 pupils need be moved to make a triangle.
In the square with 4 pupils, we need to move 1 pupil, i.e., 1 more.
In the square with 9 pupils, we need to move 3 pupils, i.e., 2 more.
In the square with 16 pupils, we need to move 6 pupils, i.e., 3 more.
In the square with 25 pupils, we need to move 10 pupils, i.e., 4 more.

We also learn that if we count the lines vertically, rather than horizontally, we will arrive at the following:

$$1 = 1$$
$$4 = 1 + 2 + 1$$
$$9 = 1 + 2 + 3 + 2 + 1$$
$$16 = 1 + 2 + 3 + 4 + 3 + 2 + 1$$
$$25 = 1 + 2 + 3 + 4 + 5 + 4 + 3 + 2 + 1$$

and that this can be written in the following way, making a much nicer figure, a number pyramid.

1 =	1	
4 =	1 + 2 + 1	3 more
9 =	1 + 2 + 3 + 2 + 1	5 more
16 =	1 + 2 + 3 + 4 + 3 + 2 + 1	7 more
25 =	1 + 2 + 3 + 4 + 5 + 4 + 3 + 2 + 1	9 more

After this we are curious to lay out the nuts according to the numbers in the number pyramid and see what shapes we get. For example taking 9 nuts and the pyramid-numbers

$1 + 2 + 3 + 2 + 1$, we get

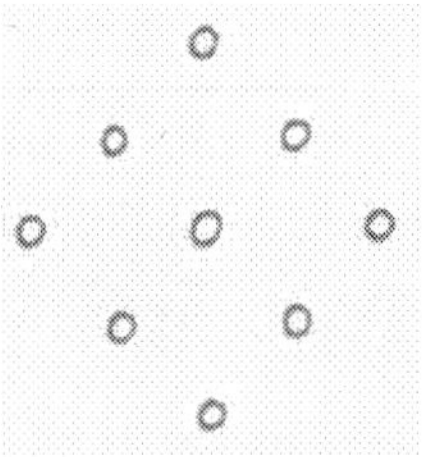

```
      o
    o   o
  o   o   o
    o   o
      o
```

and we discover that it is exactly the same figure turned up on its corner. We should have known!

Throughout the previous pages, we have worked out many sums in addition and multiplication, and in most cases, we have been "fortunate" that the numbers have worked well together. Now let's take 14 nuts and lay them out in a long line in front of us, very close together. We then take away alternate nuts and hold them in our hands. The remaining nuts now lie with larger spaces in between. We now add the nuts we are holding to the row, extending the row with the wider spacing that is now there. The 14 nuts now look as if there are twice as many.

```
ooooooooooooo
o o o o o o o o o o o o o o
```

Then we again take out alternate nuts and place them at the end of the row, spaced in the same, new way. These now take up twice as much space as before and four times as much as the very first row. There are still only 14 nuts, but because of their position they take up much more space. Eventually, if we wished, we could stretch them all the way around the world!

It is a well known fact that children up to the age of 5–6 years evaluate an object by its overall size. To overcome this stage in their development we must, among others things, let the children experience the expansion and contraction of a number of objects. By repeating such exercises they will become aware of the fact that number is not determined by size.

Back in the hall we do the same exercise with 14 children, placing them in a row. Each pupil says his number and the numbers of the 2s times table take a step forward. Easy and nice to look at! The new row makes itself more secure by linking arms, so that they can move as a unit to the end of the first row, keeping the spaces even. The original row is now twice as long.

We repeat this procedure once more, and the row is now four times as long. This is really as far as we can go, because we almost need to shout to each other now. In fact, if we stretched the row right round the world, we would have to use the telephone! Things are now beginning to move very fast indeed. We have gotten our first glimpse of raising numbers to powers.

But let us instead stay close together. The row goes back to its original size, and this time the last pupil in the row begins by running and standing in front of number 1. The next-to-last pupil then runs and stands in front of number 2, and this continues until we have two rows of the same length. The row could have been made even shorter if there were space for us to squeeze in between each other, but since we already stand shoulder to shoulder this is not possible. Had we merely been dots on a line, we could have done it—we will hear more about that another time.

In a certain way we have folded the row so that it has become 7 + 7, which is rather like folding a strip of paper. Paper also becomes twice as thick and half as long when we fold it. We could fold the paper once more, but not the row of pupils, because we are dealing with number 7. To fold once more we would need an even number.

We now try a new variation. The class stands in a row of 14 as before, and the last pupil runs forward and stands in front of number 1.

The teacher says,	"14 is ?"
and the child in the new row answers,	"1"
to which to old row replies	"plus 13."

Another pupil runs forward, and so we continue until the old row has disappeared, and the new one has taken its place. While this is going on, the conversation will sound like this:

Teacher:	"14 is ?"
New row:	"2"
Old row:	"plus 12."
Teacher:	"14 is ?"
New row:	"3"
Old row:	"plus 11," etc.
Teacher:	"14 is ?"
New row:	"14 plus 0."

For the children this is just a game of sums and quite simple, yet important for several reasons. In the first place, the loudness of the sounds from the two rows creates a sound-picture reflecting what is happening to the numbers. The new row begins in a rather shy voice, while the old row speaks with confidence. This situation slowly changes as the numbers change, until at last, the new row can shout out loudly "14 plus 0." Everyone has a chance to fight for his own row, but no one loses.

To finish this day's work, we will take 16 pupils and stand them shoulder to shoulder. We can then fold the row so many times that eventually a new row will appear at right angles to the original. This means that if we had been folding a piece of paper, we could have folded it so many times that theoretically it would come out "flat" but in the other direction!

Folding ourselves gives its own amusing result: When we are all finished, we no longer stand shoulder to shoulder, but nose to nose, in pairs—that is "flat" in the other direction.

The first vertical row shows 16 pupils all facing in the same direction. After "folding" four times, the pupils are facing one another in pairs.

What would have happened if the original row had stood looking at the backs of each other's heads, or even more amusing, if they had stood with their hands on each other's shoulders? We try these out and have in both cases prepared ourselves for the rotation of geometric figures, which we will be doing in a later year. The moment in which the columns of children are able to turn simultaneously without losing their shapes is of great importance. We can see this clearly by the way in which the children concentrate on their own positions and on that of their neighbors.

A corresponding exercise with rotation of figures can be seen on page 123. In order to stabilize the pattern, we can lock it together by asking the children to hold arms.

The next day we teach how to write out sums in a variety of ways:

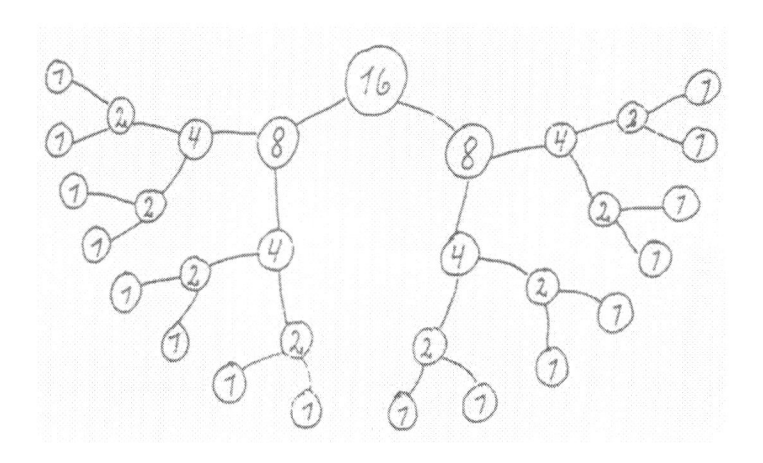

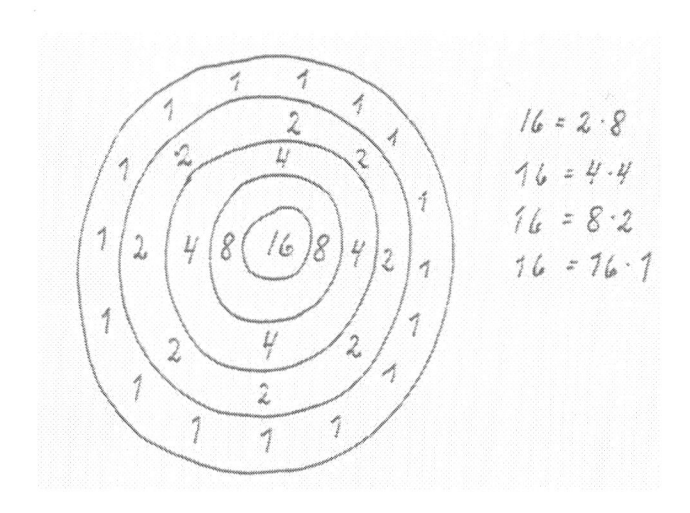

$$16 = 2 \cdot 8$$
$$16 = 4 \cdot 4$$
$$16 = 8 \cdot 2$$
$$16 = 16 \cdot 1$$

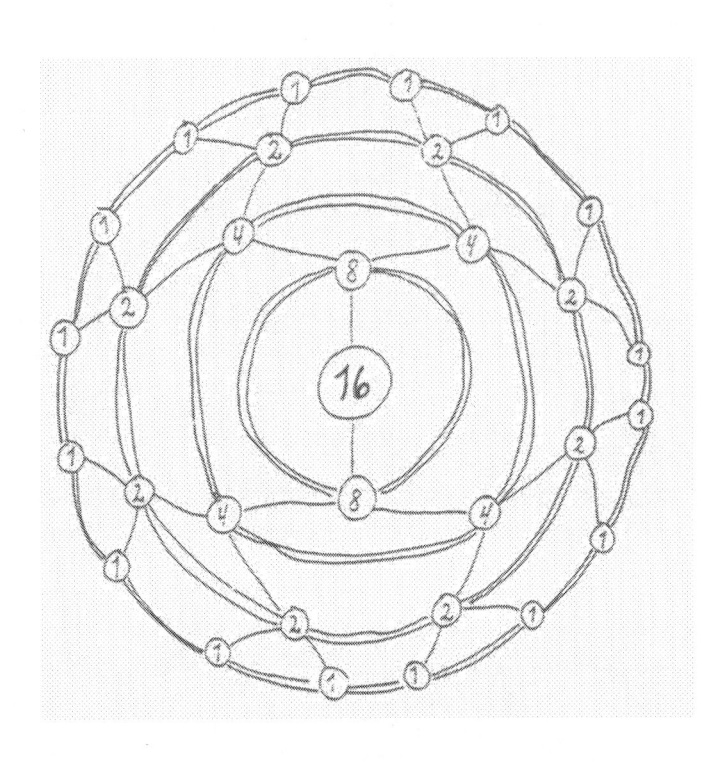

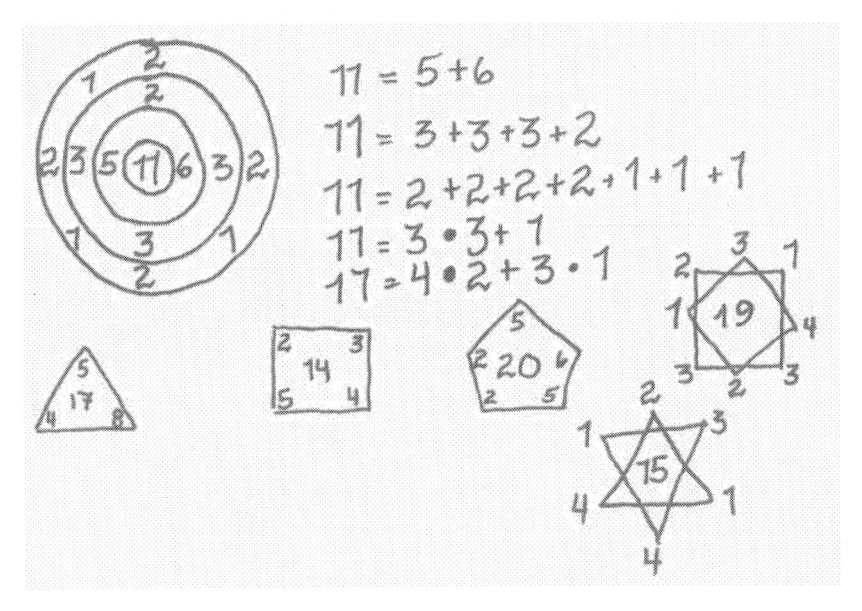

$11 = 5 + 6$
$11 = 3 + 3 + 3 + 2$
$11 = 2 + 2 + 2 + 2 + 1 + 1 + 1$
$11 = 3 \cdot 3 + 1$
$11 = 4 \cdot 2 + 3 \cdot 1$

On page 127 we had a sum which was simple, but important for several reasons. Let us look at it again:

$$14$$
$$1 + 13$$
$$2 + 12$$
$$3 + 11$$
$$4 + 10$$
$$5 + 9$$
$$\ldots$$
$$14 + 0$$

and consider the following situation:

We give the children a handful of nuts—let's say there are 20. These one child lays on his desk and is asked to divide into two piles. The choice is completely free, and he decides to put 7 in one pile. From working with the number-runner and many other situations, he knows that there must be 13 in the other pile. So he writes down,

$$20 = 7 + 13$$

He knows that the sum is correct because he can count the nuts. Now he chooses other ways of dividing up 20. He can always check the answer with the nuts if he wants to, but soon this will become unnecessary. After a few minutes he has written:

$$20$$
$$7 + 13$$
$$11 + 9$$
$$6 + 14, \text{etc.}$$

Lastly, he can put these into proper order, writing:

$$20$$
$$1 + 19$$
$$2 + 18$$
$$3 + 17$$
$$4 + 16, \text{etc.}$$

and thereby discover that he has exhausted all possible combinations. The rhythmical process of writing the sequence confirms his own conclusion.

This is the analytical approach with addition as contrasted to the synthetic approach which would let the children solve problems of the form,

$$7 + 13 = 20$$

We have repeatedly touched upon the question of the nature of the child's soul and considered it among other things in relation to phrases in the Danish government's "blue paper." If we are to put into practice the admonition, "…but first and foremost, activity and experience…," and to apply it down to the smallest detail of teaching method, then there is no doubt that the analytical method of learning addition is the one that children, given the choice, would wholeheartedly choose.

With the analytical method they use their own initiative and give play to their powers of imagination. It gives them opportunities to make decisions and experience the consequences, and this is particularly good for young children. All too often we give them decision tasks in areas where their maturity is inadequate. With this method we stretch their ability without over-stretching. Here they become more and more enthusiastic in their efforts to discover all the answers, and to know when there are no more to be found. In contrast, the synthetic method leads the child to an answer that others already know and are waiting for him to discover—knowledge is the goal. In the analytical method the goal is activity and experience itself.

Children should learn both methods, but they should learn first the analytical way, because this satisfies their primary needs. At the same time, it touches upon even deeper layers of the child's nature.

We say, "He can't see the woods for the trees" of someone who has immersed himself in detail and lost the ability to see the greater whole. Often it may be used to describe an adult who has specialized in a particular subject for many years and has come to cling on it. It is therefore, in many respects, a good characterization of our present-day society and citizenry.

A child who has been sheltered from such an upbringing might perhaps be described as one who "cannot see the trees for the woods!" The young child accepts everything with which it comes in contact as a whole and absorbs the world uncritically, whether good or bad. The child's horizon is unbelievably wide, but the view within the broad horizon has no sharpness of detail. By contrast, we adults see things within narrow limits but in sharp detail.

The child experiences the whole—for example, his father and mother, their feelings for him, for each other, and so on—but he is unable to explain what he experiences, either to himself or to others. He is therefore quite at the mercy of the whims of his environment.

Similarly, when he nears a woods, he experiences it as a whole. As he enters it, it is still a woods for him, cool and quiet, with the wind swishing softly in the gently swaying trees. An adult approaches the woods in a very different manner. Metaphorically speaking we might say that the adult takes a pair of binoculars and focuses in on a small, isolated group of trees. He notices immediately the tree stumps from a cutting and perhaps begins to count the rings to find the age of the trees. His little son, meanwhile, has taken delight in the circular rings themselves—going round and round, round and round. Therefore, remember this woods and allow the children to experience the whole first and then the parts.

Such was the method we used earlier when we spoke of the first group of numbers, "essence" numbers. We allowed ourselves to experience the whole first—for example, the starfish with its ray structure. From this we experienced the essence number five with its qualitative character. The whole gave us the number five, and fiveness was contained in the whole—in the same way as when we drew a five-pointed star inside a circle. One is therefore tempted to say that unity is the largest number, and that the "bigger" the number, the smaller it is! For qualitative, or essence, numbers, this is perfectly true.

Qualitative numbers come into existence in the world through a process of division and, within the child, through an experience of division—not through addition. It is therefore incorrect to give the children their first experience of the number five by adding unit to unit and believe that we are giving pupils in the first grade an education in harmony with their inner expectation. The following is not a picture of the number five:

It is a synthetic figure, made up of small parts.

Children have a need to unfold that activity which comes from within them and should not have forced upon them something from without. The child's number five, in this language, would look like this:

We are so used to teaching addition first because it is the easiest of the arithmetic operations, And this is perfectly correct if we are working in the synthetic way. However, if one has come to the conclusion that analysis is of the greater importance, then there is only one way to begin: by allowing children to divide in one form or another, for example, by drawing stars within a circle.

Our situation is rather like standing at the North Pole, where we can move in one direction only, south. After taking the first step, there are four possibilities and now only one of them is towards the south. We find that the same applies in arithmetic. Before we have any numbers with which to work, we must divide, and then we will see that there are four operations, of which division is one.

It is extremely interesting to remember in this regard that, historically speaking, 1 was not considered to be a number until rather recently. Not until the 15th century did the Dutch mathematician, Simon Stevin (1548–1620), prove 1 to be a number. Before that time 1 was considered to be the fountainhead from which all numbers were created, in the same way that Creation sprang from the Creator. In ancient times God's name was so holy that no one dared use it, and the same applied to the number 1. The world of numbers consequently began with the number 2, because it was the first number that arose from division of the whole. 2 was smaller than 1, lesser, because it came from 1.

When 1 was no longer the origin, the whole, but simply a number, and when the distance from 2 to 1 was no longer the distance between creator and created, then 1 itself became a number quantity among other numbers. It became necessary to specify its size with reference to a starting point, and 0 was added, with the distance between 0 and 1 as unity. The new starting point was itself a picture, with its circular form 0, symbolic of the whole.

From that moment on, 1 was no longer the greatest of all the numbers. In fact it became the smallest (of the natural numbers), and then it was valid to say: "the higher the number, the bigger it is."

When we say that it is essential for children to begin with the whole, there are two different aspects to be considered. First we consider this expression in relation to "essence" numbers and let the whole, for example represented by a circle, be divided up so that a star is formed. In this way we satisfy our need for the qualitative aspect of numbers. Secondly, and later on, we can consider the handful of nuts given to the child as the whole collection, the amount with which we begin and which must be divided up before we can do any other arithmetic at all. Through dividing we get divisions, through partitioning we get parts, and satisfy our need for analysis. In both of these exercises by moving from the whole to its components, we illustrate the cardinal nature of numbers. And in both exercises we make use of the rhythmical activity of counting, which brings in the ordinal aspect of numbers.

From analytical activity there thus arises both the elements and the foundation for doing arithmetic, and for the world of numbers itself. It is therefore of greatest importance to realize—and especially in teaching young children—that numbers do not exist in our surroundings, but only in the analyzing human mind. This is why division is the starting point in mathematics—not division as one of the four operations in arithmetic, but division as a part of the analytical study of a whole. Projecting numbers out onto our environment, for example, onto the windowpanes in the classroom (the classic mistake) makes them utterly and absolutely homeless, while we simultaneously miss observing one of mankind's greatest abilities.

Instead, give the child a lump of clay and ask him to divide it, for example, into *three*. While he is doing this, study his expression, and you will experience the ancestral home of our numbers. Similarly, let two children hold a rope taut between them, and ask a third to grip it precisely in the middle. From the assessing look in the child's eyes and the hesitation in his hand, we will again experience a number being born.

Dividing a rope in the middle demands a sure eye.

Then ask two children working together to divide the rope into three equal parts and you will be further convinced.

Threesomes have always caused problems.

Finally, ask one of the children holding the rope to swing it round and round while his partner holds the other end still. This gives them a picture of the whole.

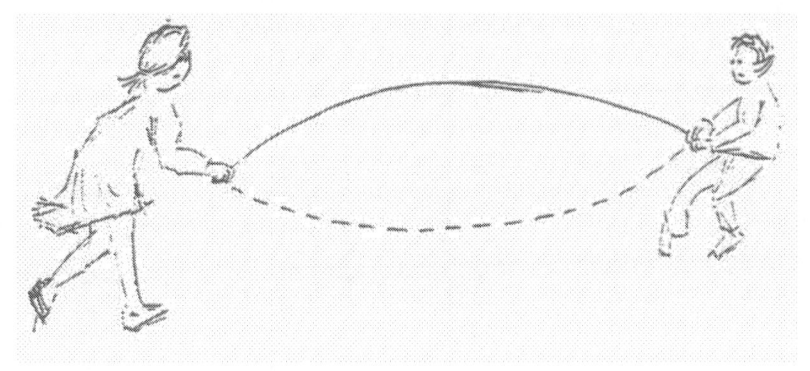

A good old-fashioned game for the first lesson in arithmetic

Then ask the child to practice swinging it this way:

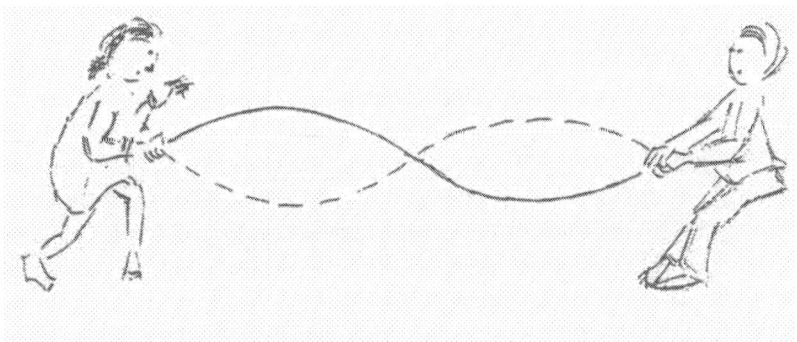

In order to do this, we may need some magic!

If we watch the child while he does this, we will see a number-producing being at the peak of activity, particularly just at the moment when he gets ready to make the first swing.

Later on in the classroom we can practice dividing the swinging rope into thirds and fourths. This demands great skill, and is in fact arithmetic "par excellence." It is also good preparation for other things—for example, the teaching of acoustics, later on.

By using exercises of these kinds we are giving the children the very best possible experiences of numbers, not through adding piece to piece, but by dividing up the whole, in one way or another.

Here is the method simplified:

and not:

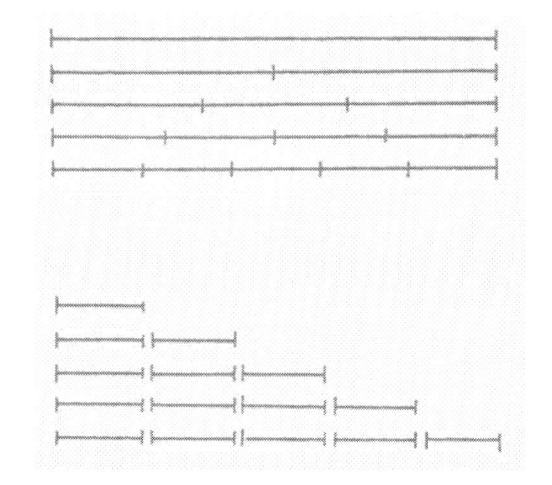

As already mentioned, another excellent starting point for teaching arithmetic is to give the child a lump of clay and ask him to divide it equally.

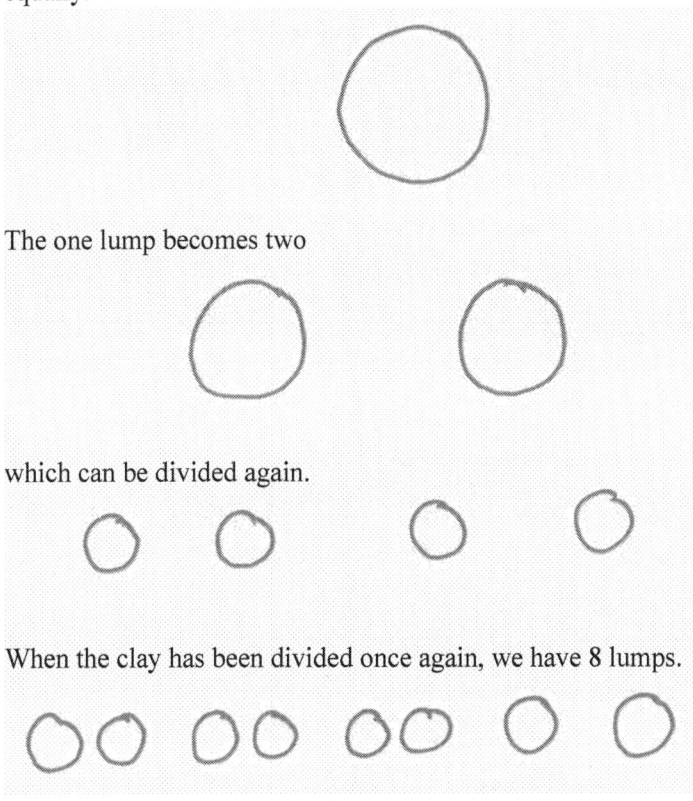

The one lump becomes two

which can be divided again.

When the clay has been divided once again, we have 8 lumps.

We can now do lots of arithmetic because our world is full of possibilities. For example we can experience that 8 = 5 + 3.

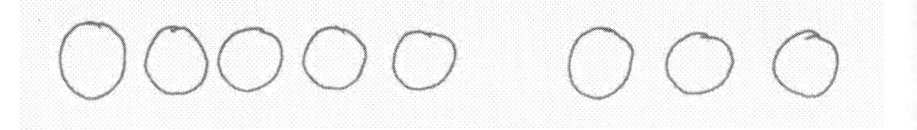

We can also hide some of the lumps so that the child can see only 2 of the 8 and ask him how many are hidden. We have thereby subtracted. Similarly through multiplication we can arrive at our number 8:

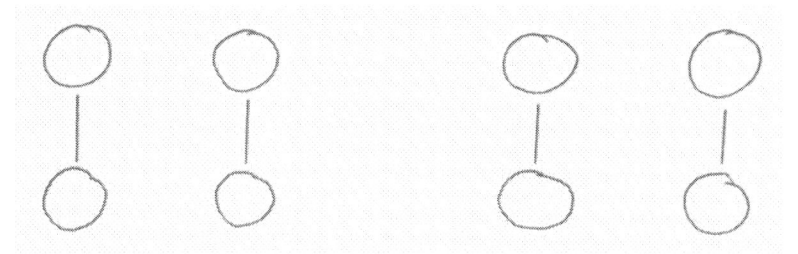

And through division, at our number 2.

We take our clay and apply a different quality to it. From our whole

we could create:

and the next time:

Each time we divide the clay, we can take a new number quality, a pre-cardinal number as the starting point, and the result will be a number with which we characterize the quantity that has come into being.

All cardinal numbers lead back to an original whole. This simple relation is the foundation of arithmetic for children of all ages. If we ask a pupil in the first grade to divide the clay into 24, we are activating a power in the child which is far above the capability of any animal. Such activity demands the use of man's spiritual powers. If we ask a pupil in the seventh grade to divide the clay into 24 pieces, we witness the same thing. The inner organ of number becomes active, and it may result in a halving of the clay three times and then a division of the 8 lumps into 3. There are other ways one might do it, each following a certain system of thought. We could even have divided our lump directly into 24 pieces, but we would then have lost our clear overview of the process.

Let us consider how we would go about dividing our clay into 19 pieces. 19 is a prime number and we experience once again the special character of these numbers. Dividing our clay into 19 parts immediately demands a great deal of us. There are no shortcuts, such as first dividing it into two.

Now try dividing it into 50 pieces! For example, halve it, then divide each piece into 5, then into 5 again.

Divide it into 28 parts. For example, halve it twice, and then each piece into 7 parts.

In this way we can work out increasingly difficult sums, but in principle, we are doing the same thing as when we told the first grade class: Divide it into two equal parts! The inner activity is experienced on different levels of difficulty, but in all cases we are starting from a whole and moving to the parts. Therefore, in different ways, all cardinal numbers lead back to 1. With the prime numbers, however, there are no shortcuts, and only one way works. Each prime is its own unique self, even the largest of them we can imagine.

All cardinal numbers lead us to 1, but where, we might ask, do the numbers we called rhythmic numbers, which are so closely related to the ordinal numbers, lead? Here we will recall the pupil, many pages back, who while walking and counting backwards, reached the number 1 and then took one more step and said, "Zero." He stepped backwards into the true starting point of the rhythmical numbers, the origin of all the tables. This is why we said earlier: "Now - 1 - 2 - 3 - 4 - 5," and then afterwards: "5 - 4 - 3 - 2 - 1 - zero."

We saw it, too, when we drew in our books as follows:

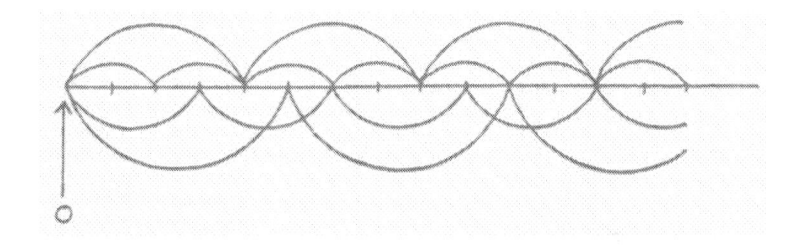

0 and 1 are therefore the points of origin, and it is from these that we begin with our ordinal and cardinal numbers. When we move backwards, we always return to the starting points, regardless of the number with which we began.

All these experiences are very much a part of daily school life for a first grade child. We cannot theorize about these things with our pupils but must simply allow these experiences to be discovered by the children through their own bodies and movement.

Later on, we can raise this theme to the level of thought. For example, when we are teaching indices and looking at the question of why

$$2^0 = 1$$
$$3^0 = 1$$
$$4^0 = 1$$
etc.

—that is, why all numbers raised to the power of 0 are equal to 1. Each and every number is closely related to both 0 and 1, and this can be clearly understood in the following expression:

$$a^0 = 1$$

Here we see the mutual relationship between "a" which stands for any number, the symbol "0" and 1. The "0" works to bring out the unity which is hidden in every number. By this reasoning, we are going in the other direction along the same path which the children in the first grade followed when they moved from the whole to the parts. By the end of their school years the pupils are able to understand this 0-problem with their thoughts. In the beginning they act it out with their hands every time they are asked to divide a lump of clay successively according to a given principle, for example, into three. When we use several different dividing principles, for example, when we seek to form the number 28, we are still starting with the whole and arriving at a cardinal number. This is our perspective now, in these exercises for the lower classes.

Later on in school life we will be dividing 28 up into its prime factors—the same thing with another *name*. The whole will then be 28, and we will work towards finding its factors.

In the first case we had a lump of clay which we grasped with our hands and divided. In the second case, we have a cardinal number, a quantity, which we grasp within our minds and try to understand its make-up by dividing it up.

While working with these relationships, we approach another problem in the teaching of arithmetic in primary school. The question is: How do we bring the physical world into the subject of arithmetic's "how do we make it real?" Yet, at the same time, how can we free ourselves of the concrete, the worldly, in order to experience the laws of mathematics in all their purity?

Here we have two completely different aspects of mathematics. On the one hand, mathematics leads us to the intangible world of logic, and on the other, it is an instrument for solving practical problems in the physical world.

Here again we must start from the child's situation and not from the adult world.

When we meet a very old person, we may have the impression, from the look in their eyes, that they are far off in another world. The same applies to the young child who also lives in a distant world, and here, too, we can see it in their eyes. But how differently they look out upon the world! The one is heading away from a long life, whereas the other is nearing life and becoming more and more our equal. These two meet as the one moves towards a physical life and the other moves away from it. And, as is often the case in meeting, they face in opposite directions, which may be one of the reasons why many children and older people have so much to say to each another. The one longs with all his will to take hold of the world, which for him is a single, undifferentiated whole. The other, with a whole life's experience behind him, speaks enthusiastically of the world's great multiplicity, interconnectedness, and complex interrelationships. The first knows little of all this, but feels within a true desire to act. The other feels that he has acted and may now leave that phase of life with the knowledge gained therefrom. Both live on a spiritual level, each with his spiritual content, which for the one consists of great fantasy pictures and for the other, of clear thoughts.

Therefore, we make the usual mistake of trying to turn a child into a small adult when we give the "ideal" arithmetic lesson and present the child with an array of different objects, hoping that he will free himself from the qualities of the objects, the objects themselves, and see only the number of "things," the cardinal number. One places the child in the physical world of objects and then immediately asks him to abstract himself away again, carrying with him only the number of objects, which exists not in the surroundings but only in the human being's productive inner life.

At the same time we expect, without a thought for the burning interest which this growing child has for everything which he meets, quite unreasonable emotional hops during an arithmetic lesson in which something totally different ought to be being practiced. We then force him or her into the laws of logic, that is, into the world where the older person lives when he has given up the material world and raised himself to the level of thought.

A child should not abstract lines of thought out of the physical world but, rather, put his line of will into the world.

Thus, we grasp the lump of clay and go into it with our hands, creating between them a multiplicity of our own. At the same time it is a multiplicity of identical objects, and we can count without being distracted. With them we can slowly transform our fantasy pictures to thought pictures, and with our newly acquired thinking and counting abilities, we can continue our way into the material world. Later on, the day will come when we can master the world mathematically, and this will be because we have consolidated our inner world of thoughts.

Counting objects is quite necessary, but with objects which the child himself has created and is not distracted by—because he is completely absorbed in the creative process and above all because the objects are identical.

Without outside interference imaginary pictures can now be transformed into thought pictures at the same time that arithmetic is being tied to something outwards, concrete, and tangible.

Arithmetic, too, has its need of physical objects when the child takes his first steps in the subject, but they should help, not hinder, the process. This is analogous to the funny situation which many parents have observed when their child is learning to take his or her first real steps—the child has long been able to stand up with the help of a table or mother's hand, but not yet dared venture out into deep water. But one day he is there, standing with a hairbrush in his hand—and with that invaluable support, which the hand senses and feels just as fully as the solid table leg—he now balances right across the room and to the safety of his father's trouser leg.

The brush was the necessary, fixed point in a difficult physical world. It was light, did not get in the way, and with it in hand the child learned to walk. It was not carried with raised arms to be given to father but, rather, as a needed support during the walking movement.

In the same way, the lump of clay (or pile of nuts or pile of beans or whatever) is not a burden but simply a necessary physical satisfaction for the senses, so that movements of thought will be balanced while the art of arithmetic is being practiced.

Later on we can involve different objects without causing distractions, remembering that they need not be different purely for the sake of abstraction, but rather as they arise naturally out of circumstances.

Eventually the brush becomes obsolete, but we still pick it up and walk around, happy in the knowledge that we do not really need it. The same can be said for our lump of clay. We can turn to it when necessary, but at the same time we know that we are capable of using pure numbers and can experience that freedom. This is important, and we can do it all without losing ourselves in the abstract.

In the examples above we have shown a number of ways to work with pure numbers through the use of the picture element rather than the abstract. We use such exercises often during the transition from counting exercises to those based upon cardinal numbers. Number squares are another exercise belonging to this category. A classic example is from Dürer's etching *Melancholia*.

A number square can be made in the following way. First we divide a square into its smaller squares and put the numbers 1–16 into them as below:

1	2	3	4
5	6	7	8
9	10	11	12
13	14	15	16

Now we exchange numbers according to the following principle:

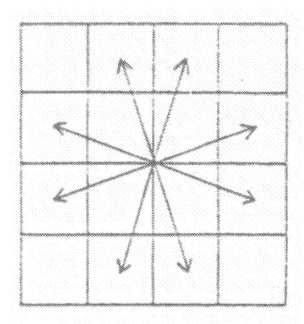

Which gives us:

1	15	14	4
12	6	7	9
8	10	11	5
13	3	2	16

If we add up the squares either by rows or by columns, we always arrive at 34. Diagonal additions give the same result. All pairs of numbers symmetrical about the middle point add up to 17, and one can therefore create a long list of combinations which add up to 34 with the help of the corner numbers. For example,

$$12 - 8 - 5 - 9$$
$$12 - 15 - 5 - 2$$
$$1 - 10 - 16 - 7$$

149

In Dürer's painting the number square is slightly different. It is created according to the following scheme:

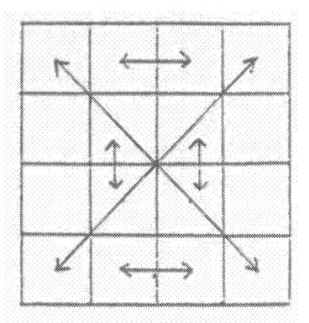

and looks like this:

16	3	2	13
5	10	11	8
9	6	7	12
4	15	14	1

The full title for Dürer's etching, by the way, is actually *Melancholia I*, where the *I* means "to turn away from." This full title gives a completely different interpretation than the short title, and it corresponds better to the picture's content. For example, the sun and the beautifully arched rainbow hardly indicate great melancholia.

Returning to number squares, another number square can be constructed in the following way:

3	16	9	22	15
20	8	21	14	2
7	25	13	1	19
24	12	5	18	6
11	4	17	10	23

In the diagram on the left, the figures within the middle 5 x 5 square remain unchanged. The numbers outside are brought into the square as shown to the right.

In this square all the horizontal and vertical rows add up to 65, as do the diagonals. Here the sum of the four corners is 52, not 65, because we have taken only four numbers and are missing the fifth. We find it in the center of the number square. Now the children can find many different four-number squares whose corners add up to 65 when the 13 in the middle is included.

The children experience great delight when they discover these combinations, and it also strengthens their ability to work with numbers.

Other examples of arithmetic with pure numbers, for the transition from rhythm to quantity, are given in the following. The first can be practiced very early on in a simple form, and we will return to it much later in connection with the formula for the sum of a series and for looking at difference sequences. It can be varied endlessly with shorter rows or longer rows, by making the number of elements even or odd, etc.

The task is to add up the numbers from 1 to 10. We do it this way:

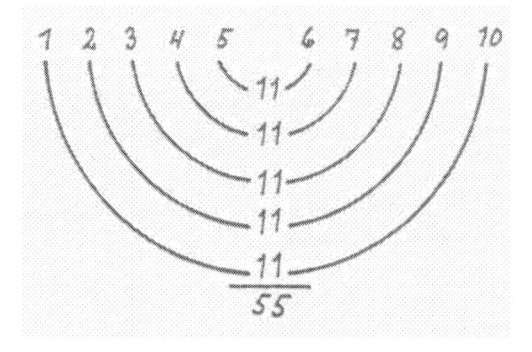

```
1  2  3  4  5    6  7  8  9  10
                11
               11
              11
             11
            11
            55
```

We could also have placed a zero in front, as shown below, thus displacing the 5, but the sum does not change:

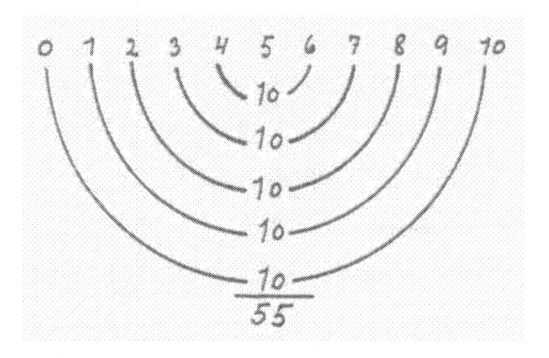

```
0  1  2  3  4  5  6  7  8  9  10
                10
               10
              10
             10
            10
            55
```

From the 10 we could take 5 and place it where the 0 is so that we have 5 at the beginning and 5 at the end. Then we could take 4 from the 9 and give it to the 11 so that we would have 5s where the 1 and the 9 are. If we were to continue in this way, the old row of numbers,

$$0 + 1 + 2 + 3 + 4 + 5 + 6 + 7 + 8 + 9 + 10$$

would become:

$$5 + 5 + 5 + 5 + 5 + 5 + 5 + 5 + 5 + 5 + 5 = 11 \times 5 = 55$$

It is a very fine exercise for the children to try to even out the row and make all the numbers equally large in this way. With an odd number of elements to add we always have a middle number, and where the number is even, we can always add a zero to make it odd. With this basic form we can now develop many examples relating to the tables.

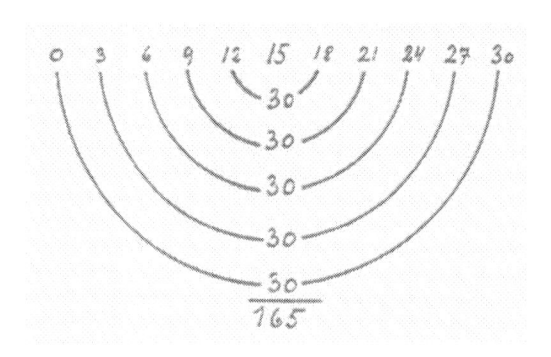

```
 0   3   6   9  12  15  18  21  24  27  30
                     30
                    30
                   30
                  30
                 30
                ─────
                 165
```

Given an irregular series of numbers such as:

    1   4   5   7   10   12   14   19

we have to feel our way forward. We might begin by supposing that all the numbers can be leveled out to tens, and with this in mind we take 9 from 19 and add it to the 1, so that we have:

    10   4   5   7   10   12   14   10

Now we could take 2 from 12 and 4 from 14, and add these 6 to the 4 to make another 10:

    10   10   5   7   10   10   10   10

We now see that it does not work out with 10, so we try 9. The number 5 has to have 4 added to it, which we will take from the last four 10s, giving:

10 10 9 7 9 9 9 9

and finally:

9 9 9 9 9 9 9 9 9

Such exercises cannot always be solved with a whole number. So we do the best we can and wait for a later lesson to learn the way to find a more satisfying answer. For example, if we have the following numbers:

5 7 8 11 17 18 20

we think of the number 11 because it is in the middle number. We make the following changes:

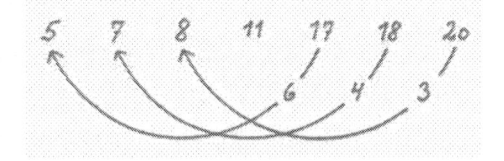

5 7 8 11 17 18 20
                6   4   3

and arrive at:

11 11 11 11 11 14 17

Obviously 11 does not work, so we try 12 by taking 6 from 17 and sharing it amongst the 11s. This gives:

12 12 12 12 12 14 12

We now realize that we cannot do any better, because there are 2 too many in one of the numbers.

There will always be children in the class who find it difficult to distribute numbers this way, And it may be necessary to use nuts as an introduction to these exercises. For example, one can ask them to lay out 4 nuts in one row, 5 nuts in another row and 6 in a third row. Then, we ask them to redistribute the nuts so that all rows have the same number of nuts. Only one nut need be moved.

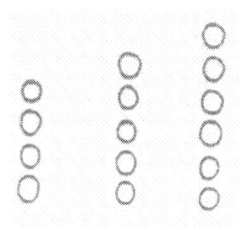

It is important for them to realize that one move is enough to make three rows equal. Slowly and surely, we now make the task more and more difficult.

In the last task of the exercise illustrated above we had a remainder to work with. The following exercise is related, and we simultaneously use averages and practice estimating, an ability which is very important to develop.

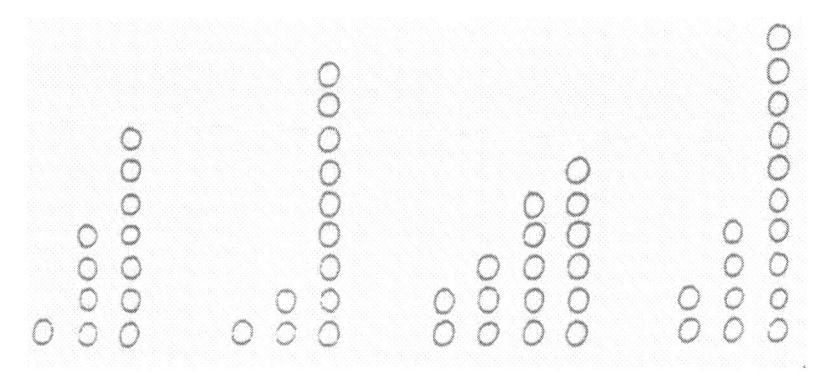

155

We give 11 children a specified number of nuts, for example, 10 nuts each. They seat themselves around a table, and 10 of the children take some of their nuts into one hand, keeping the rest in the other hand under the table. They place the nuts they have taken on the table, and they keep them covered with their hand. On a sign given by the eleventh child, they all take away their hands just long enough for him to glance round from one pile of nuts to the other. Then they cover their piles again and now he must try to estimate! How many nuts are there hidden in hands and on the table? We teach him to estimate the average number of nuts held by each child, to multiply this by 10, and thereby arrive at a very specific estimate of the total. We then check. Finally, we can ask the eleventh pupil how many piles of nuts he can remember exactly, and for those he remembers, how many nuts must therefore remain in the other hand under the tables. Then it is the next child's turn.

The ability to estimate an amount and then check one's answer is very important throughout a child's school life. By teaching this we can avoid many absurd answers.

The following little game is very useful for this purpose. The teacher has some nuts in his or her hand—let us say 24 nuts. One child is asked to give them out to three other children, so that each has, if possible, the same number. He is allowed only a quick look at the total in the teacher's hand, then suggests perhaps that 6 nuts should be given to the first child. The teacher hands them over, and the pupil is allowed a second quick glance to judge the remainder in the teacher's hand and to decide if the second child should also get 6 nuts. He decides not, 10 is better. The second child is given 10 nuts, and the remaining nuts are given to the third child.

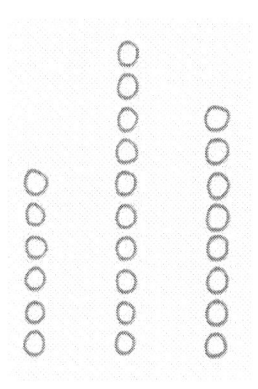

This doesn't look right, so he quickly moves 2 nuts from the pile of 10 over to the pile of 6 nuts. The teacher then collects all the nuts in his palm again, and the children survey them with their eyes and say to themselves: This is what 24 nuts looks like.

Another time the nuts can be spread out on the tabletop and covered with a handkerchief. The children are allowed a quick look and asked to estimate the number.

With all such games the children will be full of ideas for variations and improvements. Many will be excellent, but some not so useful. It is a good idea to let them try them all out—even the less useful or incorrect—as this helps the class develop the ability to make their ideas workable in practice.

Now let us make 10 dunce caps, numbered from 1 to 10, with the numbers written all around the caps so they can be seen from everywhere. 10 children are given caps, and they form a circle. The teacher calls out a number, let us say 7, and the pupils must now get together in pairs, so that their numbers add up to the number given. Those pairs then move over by the wall. The rest try to arrive at the number required by using both addition and subtraction. A subtrahend sits down next to the addendums who remain standing. When complete, the whole group moves to the wall. Is it possible to get everybody over to the walls?

Some of the children are a little embarrassed, others just the opposite, but eventually everything works out.

Thus 3 soon finds 4, 2 and 5 embrace each other, and at last 1 and 6 get together. The next round, however, will take longer, but with a little help they can find the following combinations $8 + 9 - 10 = 7$. All that is left now is number 7, who knew from the start that he or she would be left standing all alone. Now we will see if we can bring 7 to the wall as well. To do this, we must use some of the group already at the wall, and in all likelihood, use more than two of them. One possible solution would be:

$$7 + 5 - 3 - 2 = 7$$

Eventually the children will ask the obvious question themselves: Can the whole class go to the wall as one large group adding up to 7 all at once?

This problem, too, can be solved. We look at the following figure which by now is quite familiar to all, and reason that all the numbers except 7 must cancel each other out (that is, add up to zero).

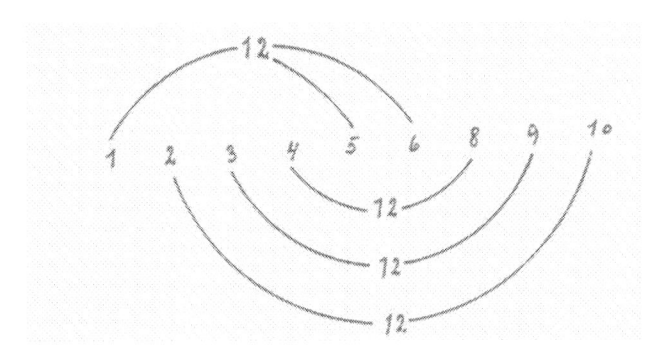

The number 12 can be made four times, and these can cancel out.

The whole problem could be solved quite simply in another way—if we had had an eleventh hat with 0 on it, since 7 + 0 = 7.

Now the 10 pupils are standing in a ring again, and we ask them to multiply themselves with each other so that the results are numbers between 10 and 20.

10 immediately selects either 1 or 2, knowing this is the only way he can join in, but 9 is too quick for him and has taken 2 because that is his only chance. This leaves 1 for number 10 to take, which satisfies number 1 for obvious reasons! 2 is not at all sure whom to choose, and before he can come to a decision, 9 has taken him by surprise. 7 and 8 suddenly feel unwanted, although they had been very active at first. 3 reacts very differently; he remains calm, knowing that somebody is eventually going to need him. He knows only too well why 6 is so keen to get to know him, and he helps 4 and 5 to understand that they are perfectly all right together.

This game can be varied in many ways, also using subtraction and division.

For example, the teacher writes the numbers 1 to 10 in a circle on the floor. Then the children wearing hats go and stand on their respective numbers. The teacher now writes a number on the blackboard, for example 8, and asks the ten pupils to move to a number on the floor which, when added to their own number, makes 8. Some of the class must leave the circle because they cannot achieve the number. The whole class will be able to check the results, knowing immediately why three of the places in the circle are unusable, and why number 4 stands so proudly in the same place.

If we use subtraction, we run into some complications: should the hat be subtracted from the floor, or the floor from the hat? Perhaps it is best to allow both possibilities. Either way, many of the class will have to leave the circle, and even more when we use multiplication and division.

We now need to make a 10 x 10 square on the floor for the numbers 1–100, in one way or another. A tile floor with 50 x 50 centimeter (18 x 18 inch) tiles would be perfect. We could also get by simply marking out small crosses where the squares are to be and putting in some of the numbers, for example, as shown in the illustration below. The children will soon be able to find their way around to the various numbers.

1		5		10
11		15		20
21		25		30
31		35		40
41		45		50
51		55		60
61		65		70
71		75		80
81		85		90
91		95		100

The teacher might then say to the children, who stand ready with their dunce caps on: "Go, stand on your numbers!"

The pupils stand now on the first 10 places.

"Add 5 to your hat numbers!"

They now stand from 6 to 15, the row has been broken up, and 1 and 10 are standing near each other.

"Add 15 to your number!"

The pattern is almost the same.

"Add 22 to your number!"

The row is broken up again, but in a different place.

"Take 4 away from your number!"

Some of the children are now out of the game.

"Go back to your original place, and when I say 'go,' multiply your number by 2 and move to the new place!"

The rows from 1 to 10 now change to 2 rows with a double space between them.

"Multiply your number by 3!"

Now we see a pattern emerging, which we may remember from earlier and which we might later write down in our books.

"Multiply by 4!"

"Multiply by 5!"

The pattern is a very special pattern now. Etc.

"Divide your number by 2!"

A short row with numbers 1–5 appears and 5 pupils must go out.

"Divide by 3, by 4," and so on.

Now there are not many children left in the square.

"Divide 60 by your number!" (This must have been carefully prepared earlier).

Now the children are widely placed.

A floorboard containing a hundred squares ought to be a permanent feature of the school gymnasium because it can be a great help in arithmetic. But it is also possible to do some of these exercises using our number-line. For example, try multiplying by 1, at 3, at 4, etc., and see the resulting movements and final patterns.

The same exercises can be done in yet another way. Ask the children to bring pocket flashlights to school one day, preferably ones with-sharp conical beams. The numbers from 1 to 100 are written in a square on the blackboard instead of on the floor, We darken the room and let 10 pupils light up their white chalk numbers with their torches. As we give various instructions, for example, "Multiply your number by 3," we see the patterns emerge on the board.

Here is yet another exercise for the transition from rhythm to quantity numbers, one which gives lots of practice in doing sums:

"Add together alternate numbers along the row!"

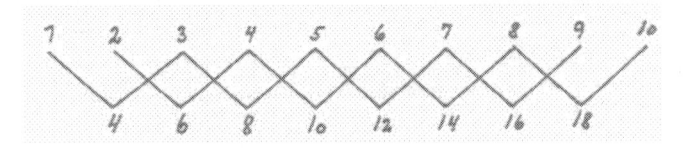

```
 1   2   3   4   5   6   7   8   9   10
   ╲ ╱ ╲ ╱ ╲ ╱ ╲ ╱ ╲ ╱ ╲ ╱ ╲ ╱ ╲ ╱ ╲ ╱
    4   6   8   10  12  14  16  18
```

162

"Add every third number, in pairs."

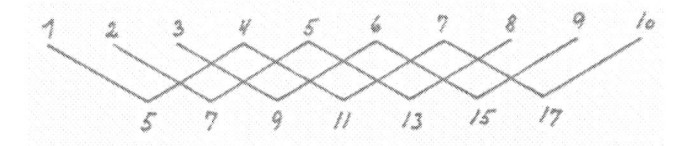

```
 1   2   3   4   5   6   7   8   9   10
      5   7   9   11  13  15  17
```

"Multiply alternate numbers with each other!"

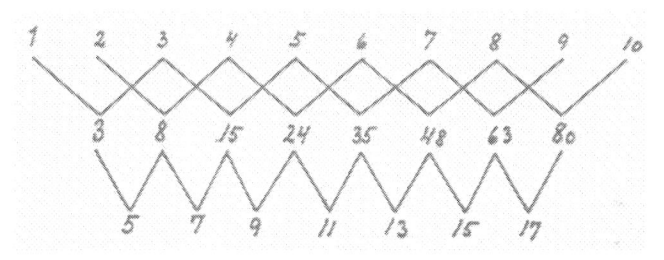

```
 1   2   3   4   5   6   7   8   9   10
      3   8  15  24  35  48  63  80
      5   7   9  11  13  15  17
```

As soon as the children see the pattern of the sequence in a solution, they are ready for the next problem.

Taking the natural numbers in sequence three at a time calculates that which corresponds to (1 x 3) + 2 for the first three numbers 1, 2 and 3.

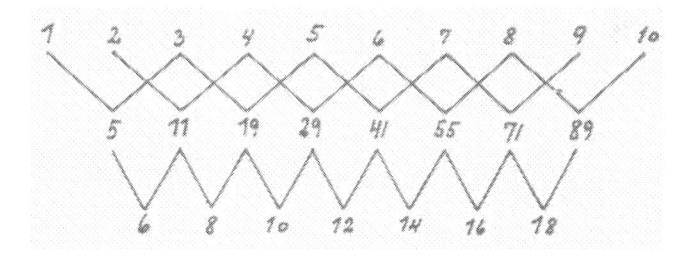

```
 1   2   3   4   5   6   7   8   9   10
      5  11  19  29  41  55  71  89
      6   8  10  12  14  16  18
```

163

In the same way, work out that which corresponds to 1 x 2 x 3.

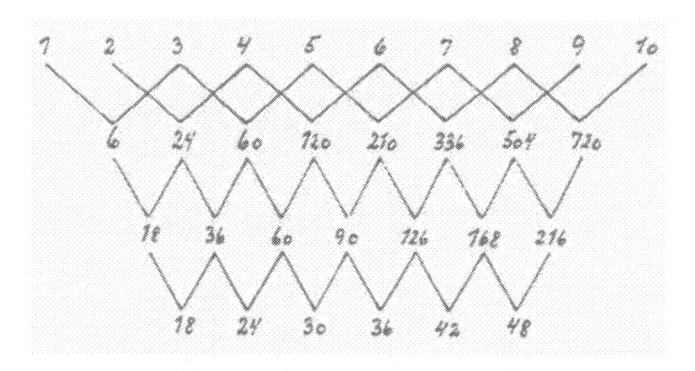

```
  1    2    3    4    5    6    7    8    9   10
       6   24   60  120  210  336  504  720
           18   36   60   90  126  168  216
                18   24   30   36   42   48
```

The pupils can vary these exercises themselves and perhaps come upon the following two "different" sums:

        (3 x 2) / 1, (4 x 3) / 2, (5 x 4) / 3, etc.
        and (1 x 2) + 3, (2 x 3) + 4, (3 x 4) + 5, etc.

Later on let them make the exercises more difficult, perhaps like these:

        (1 x 4) + 2 + 3, (2 x 5) + 3 + 4, etc.
        (1 x 3) + (2 x 4), (2 x 4) + (3 x 5), etc.
        1 x 3 x 4, 2 x 4 x 5, etc.
        1 x 2 x 4, 2 x 3 x 5, etc.

Chapter 9
CONCLUDING REMARKS ON MEMORY, PLAY AND MATHEMATICS

In the preceding pages an attempt has been made to describe three types of number experience. As adults we experience these to varying degrees, with cardinal numbers making the greatest impact because we use them constantly in our daily lives. The rhythmical process, which actually is the basis for the ordinal numbers, is of lesser importance to us. As for "essence" numbers, we experience them only in isolated moments, for example, when we take a close look at our emotional state, are enchanted by the beauty of the plant world, or look with wonder upon the geometrical forms of the mineral world.

For children this situation is reversed. A large part of their day is spent in rhythmical movement. We can see this when we observe a young child playing or perhaps going for a walk with his parents. As for essence numbers, the child identifies with them through fairy stories, assimilating the pictorial relationships and the qualities depicted in the adventures. The child's understanding of cardinal numbers, on the other hand, is built upon an unclear experience of one versus more than one, or many.

The child is on his way into our world, and we must accompany him on the journey, supporting him with our teaching. Teaching is directed towards the soul in the same way that food is directed towards the body.

Our teaching is somewhat intangible, while the food we give the child to eat along the journey is of a more material nature. The soul of the child is closely connected to the body, and the body is ruled by rhythmical processes, and so must we characterize our teaching in the first years. The moment food is inside the body, it is influenced by rhythm. The process commences with preparation by the teeth and continues through the whole of the alimentary canal. In the same way, our teaching must have a rhythmical character, and on the previous pages we have given many examples of this for use in the younger classes.

Throughout these first years the connection between soul and body is gradually loosened. By the time of transition from grade 3 to grade 4, powers of memory and understanding have become much more accessible than when school first began.

The ability to learn by heart grows, and this has significance in teaching arithmetic. We have said earlier that learning by heart is essential but that the method used is also of importance, and it is this aspect which we will deal with on the following pages.

When we study the activity of memory, it becomes obvious that in the past people have had an inaccurate concept of how memory images appear in the human soul. Many descriptions of these phenomena are based upon the belief that at the original moment of observation, we attach an image to our sensory perceptions. When the perceptions have ceased, the image is stored in the soul, so that at the moment of recollection it can be recalled from the soul and placed anew in our consciousness.

All self-observation shows, however, that memory images are not just stored within us complete and ready, waiting to be recollected, but are created anew in every instant, each time we look at them. Old images do not exist, no more than old music, which also has to be recreated. We might think that music exists on a record, but it does not exist directly, only indirectly, for the groove exists and gives rise to the music, but is not the music itself. The music is recreated by playing the record each time we hear it. With live music there are usually notes which are the permanent feature, but here too the music itself is created anew each time.

We may now ask, what is the permanent feature in our memory images? A study of our memories through life, and their intensity and character in the different stages of our lives, points with all clarity to the foundation on which they rest. Why do we all remember our early childhood experiences so well? Why are they preserved for us in such scope and strength, as no others? What accompanies later experiences, which also are imprinted in our memories? What is meant, in fact, by the expression "imprinted upon memory"?

In other words, what is the sphere of life which is so susceptible to impression, that we are able to recall? The life situation of children shows us very clearly that our ability to remember is completely dependent upon our action-movement and emotional lives, to which the rhythmical processes are connected. These rhythmical processes strongly dominate our early years. We become more and more cut off from them as time goes on, but also more free and independent of them.

All of our experience affects our will and our feelings. If we could continuously check our pulse and breathing, we would see that here we have a sensitive barometer for measuring our state of mind. Here are our experiences imprinted—not as images, but as marks upon the soul, which become visible for our inward observation. Inward observation needs, in the same way as outward observation through the senses, a supplement from imagination in order to be recognizable. In the case of memory we recognize the images which our imaginations supplement us with—

we have seen them before—and we call them memory images. They are, however, also new, and exist only at the moment of recall. It is the emotional impression upon the soul which is permanent within the whole process.

When we therefore ask what it is which is permanent in memory, we have to go much deeper than the life of thoughts. Thus, in teaching younger children, it is the worst imaginable method to tell pupils to go home and learn something by heart for the next day. This would in fact mean—go home and give your inner soul, where rhythm and feelings are a part, such impressions as are needed for you to create memory images tomorrow. This is simply too much to ask of them.

It is better instead to practice the multiplication tables games before the end of the school day. Here we are many children all at once, doing the same thing. Here we can help each other make things work out right and help each other discover the many interesting relationships. Here we move around on a big floor in a big room, here we experience excitement and the multitude of possible combinations. In short here we have all the elements necessary to make impressions, so that there will be something to observe on the inner blackboard tomorrow.

It follows from this, therefore, that the multiplication tables provide us with the best imaginable opportunity for exercising pure memory. In the first place, we are concerned here with pure numbers and work thereby on a level where we are not affected by the outside world of objects. Secondly, we have the opportunity to immerse ourselves in rhythmical relationships, and thereby find ourselves in precisely that area which alone can create a foundation for memory images. Therefore, when rhythm is properly incorporated and appreciated, the multiplication tables become the first place where pure memory can be exercised with children.

After a good foundation has been laid using rhythmical exercises such as these, in any particular area, we can then give homework. Homework in the younger classes which does not have strong impressions as a basis can hardly be expected to be useful. We ought therefore always ask ourselves before giving any homework, if we have ensured that lasting impressions have been made upon the child's inner life, for example, using stories, rhythmical movement, or nuts laid out on the school desk, divided into piles of different sizes, shaped into squares and triangles, and so on. If we have done this, then we can be sure that the impressions, of themselves, will be waiting there and longing to rise up into consciousness.

This may happen the following day in school, and we can prepare for it by giving homework—homework which is directed to a point within the child lying halfway between its feeling-world and its thought-world, or in other words, homework which has a musical element within it.

For homework, too, should be based upon what the child is, and not upon what we wish it to be. It is much too easy with homework to aim at "imparting basic knowledge" and very difficult to nourish the child's will and emotional self. There is, however, no other method. For the child who is 8–9 years old in the morning classroom is the same age later in the day when he does his homework. He has the need for activity and experience all day long—and in arithmetic, too.

For this child is on his or her way into our world—a metaphor we used before in the example of children meeting older people who were moving away from their long lives. Let us imagine that some of these older people were experienced mathematicians and we could listen to their conversations with the young. This would be interesting, for the two would have important, common experiences to discuss, since the distance between a child's play and the most difficult, most advanced mathematics, as has often been pointed out, is short. The child's play is in one way raised above the physical world, just as mathematics also is. In his play the child uses objects of this world but frees himself from the normal laws and creates his own rules. His objects must obey entirely new rules, sprung out of the child's own fantasy and desire to act, and they enter into entirely different relationships with each other than those that prevail when the child turns his back and leaves them to themselves.

In the world of the great mathematicians—not necessarily only the old—we find a similar situation. Their world of thoughts is raised above the earthly, and when they do make use of earthly objects, it is only as a pictorial representation for something else. A line or circle on a piece of paper is only a reference to an ideal line or an invisible, perfect circle, which exists only on a spiritual level. A tangent to a circle, drawn on the blackboard, has infinitely many points in common with the circle, but in the world of ideas they touch at only one point. It has been pointed out over and over again the important role which fantasy and imagination play in the world of mathematics and how we there may create without limit new forms of mathematics—new forms which find support not in adult thinking alone, but also build upon activities more reminiscent of children's intuition.

Play and mathematics belong together. They appear beside each other as related but nonetheless quite different phenomena, like the child and the elder from the same family, as they meet—related but each one facing in his own direction.

And by "the elder" here we mean in pictorial language not so much the mathematics which is studied daily everywhere, but rather the highest forms of mathematics. It is really here that mathematics can be compared to the play of children.

This is important for another reason. For we have often seen that the great creative advances in mathematics are not connected with older but instead younger, sometimes very young, mathematicians (Abel, Gauss, Pascal and others). They are yet in possession of the inspiration of youth and still affected by precisely those senses which form the basis of mathematical talent. The great works of mathematics were not come upon through pedantically worked-out proofs to begin with, but through a spontaneous, intuitive experience of unity and completeness. The proof, the logical thinking to show to others, came later. These great creators have followed the same path as the child, only in the opposite direction, and they have also relied upon experiences of the same sort—such as a specially developed talent for remembering experiences from their earliest childhood, in connection with the sense of balance, the sense of movement, and so on.

Therefore, when we look for the really vital elements to use in teaching mathematics to young children, we are helped both by studying the laws of child development and also by studying the lives of the great mathematicians.

Made in the USA
Middletown, DE
15 November 2018